THE ECONOMICS OF URANIUM

THE ECONOMICS OF URANIUM

Anthony David Owen

PRAEGER SPECIAL STUDIES • PRAEGER SCIENTIFIC

New York • Philadelphia • Eastbourne, UK
Toronto • Hong Kong • Tokyo • Sydney

Library of Congress Cataloging in Publication Data

Owen, Anthony David.
The economics of uranium.

Bibliography: p.
Includes index.
1. Uranium industry—United States.
2. Uranium industry. I. Title.
HD9539.U72U533 1985 338.2'74932 85-3583
ISBN 0-03-003799-9 (alk. paper)

Published in 1985 by Praeger Publishers
CBS Educational and Professional Publishing, a Division of CBS Inc.
521 Fifth Avenue, New York, NY 10175 USA

56789 052 987654321

Printed in the United States of America on acid-free paper

INTERNATIONAL OFFICES

Orders from outside the United States should be sent to the appropriate address listed below. Orders from areas not listed below should be placed through CBS International Publishing, 383 Madison Ave., New York, NY 10175 USA

Australia, New Zealand
Holt Saunders, Pty, Ltd., 9 Waltham St., Artarmon, N.S.W. 2064, Sydney, Australia

Canada
Holt, Rinehart & Winston of Canada, 55 Horner Ave., Toronto, Ontario, Canada M8Z 4X6

Europe, the Middle East, & Africa
Holt Saunders, Ltd., 1 St. Anne's Road, Eastbourne, East Sussex, England BN21 3UN

Japan
Holt Saunders, Ltd., Ichibancho Central Building, 22-1 Ichibancho, 3rd Floor, Chiyodaku, Tokyo, Japan

Hong Kong, Southeast Asia
Holt Saunders Asia, Ltd., 10 Fl, Intercontinental Plaza, 94 Granville Road, Tsim Sha Tsui East, Kowloon, Hong Kong

Manuscript submissions should be sent to the Editorial Director, Praeger Publishers, 521 Fifth Avenue, New York, NY 10175 USA

For Jackie, Catherine, and Rhys

Acknowledgments

I would like to thank Ian Duncan, Jim Munro, Keith Robb, George White, Jr., and Ron Wilmshurst for the considerable amount of time and patience they expended in reading earlier versions of this manuscript. Their detailed comments and criticisms are gratefully acknowledged. Neither they, nor the organizations with which they are associated, are responsible for the views expressed herein or for any remaining errors of fact or interpretation. I also wish to thank Barbara Flynn and Silvana Tomasiello for typing various drafts of the manuscript.

The assistance provided by the Organisation for Economic Cooperation and Development, *Nuclear Engineering International*, and NUEXCO by kindly giving permission for the use of copyright material is also gratefully acknowledged.

Finally, my thanks must go to the University of New South Wales for granting me a sabbatical leave in 1981, which allowed me to lay the foundations for this book.

Contents

List of Tables

List of Figures

List of Abbreviations

AEC	Atomic Energy Commission (U.S.)
AGR	advanced gas-cooled reactor
bbl	billion U.S. barrels (of oil)
BWR	boiling water reactor
CANDU	Canadian Deuterium Uranium
CDA	Combined Development Agency
CEA	Commissariat à l'Energie Atomique
CFP	Compagnie Française des Pétroles
COMUF	Compagnie des Mines d'Uranium de Francoville
CPE	Centrally Planned Economies
DOE	Department of Energy (U.S.)
EAR	estimated additional resources
EEC	European Economic Community
ENEA	European Nuclear Energy Agency
ERDA	Energy Research and Development Administration
FBR	fast breeder reactor
FGD	flue gas desulfurisation
FRG	Federal Republic of Germany
GCR	gas-cooled reactor
GDR	German Democratic Republic
GENCOR	General Mining Union Corporation
GWe	gigawatt electric
HTGR	high-temperature gas-cooled reactor
HWR	heavy water reactor
IAEA	International Atomic Energy Agency
IBA	International Bauxite Association
INFCE	International Nuclear Fuel Cycle Evaluation
IUREP	International Uranium Resource Evaluation Project
K cal	one thousand calories
kg	kilogram (one thousand grams)
Kw	kilowatt
Kwh	kilowatt hour
lb	pound (453.592 grams)
LMFBR	liquid metal fast breeder reactor
LTFC	long-term fixed-commitment contract
LWR	light water reactor
MWe	megawatt electric

NEA	Nuclear Energy Agency
NUFCOR	Nuclear Fuels of South Africa
OECD	Organisation for Economic Cooperation and Development
Onarem	Office National des Ressources Minière
OPEC	Organisation of Petroleum Exporting Countries
PHWR	pressurized heavy water reactor
PWR	pressurized water reactor
PUK	Péchiney-Ugine-Kulmann
RAR	reasonably assured resources
RTZ	Rio Tinto Zinc
SGHWR	steam-generating heavy water reactor
SNEA	Société Nationale Elf-Aquitaine
SR	speculative resources
SWU	separative work unit
t	tonne
toe	tonne of oil equivalent
WOCA	World Outside the Centrally planned economies (CPE) Area
WPNFCR	Working Party on Nuclear Fuel Cycle Requirements

THE ECONOMICS OF URANIUM

1

Introduction

HISTORICAL BACKGROUND

On the evening of September 24, 1789, Martin Heinrich Klaproth,[1] a Berlin chemist, announced to the Royal Prussian Academy of Science his latest discovery—a new element, which he had discovered in a sample of pitchblende. Pitchblende had been mined as early as 1517 at St. Joachimsthal in Bohemia (now Czechoslovakia) and was generally known as a compound of zinc, tungsten, and iron, but Klaproth claimed that his discovery was not related to these three metals. He named this new element "uranium" in honor of the discovery eight years previously of a new planet, Uranus, by Sir William Herschel. What Klaproth failed to notice, however, was that the material available to him was not elemental uranium but rather an oxidized form. The preparation of elemental uranium was first accomplished by a French chemist, Eugène Péligot, in 1841.

Until the end of the nineteenth century, uranium was regarded as a mineral of only minor importance; however, many of its physical properties were measured. In 1896, Henri Antoine Becquerel discovered that uranium gave off penetrating rays, a phenomenon that one of his students, Marie Curie, investigated for her doctoral thesis and named radioactivity. This discovery eventually led to the isolation of radium salts from uranium ore by Pierre and Marie Curie in 1898. Radium is always present in uranium ores in the proportion of one gram of radium to three tons of uranium.

Even though uranium salts were known to be radioactive, the primary interest in uranium and its ores until the 1930s was centered on the recovery of radium—uranium was an unrequired by-product. Uranium's major uses during this period were as a pigment for coloring glass and painting china, and as a substitute for tungsten.

Toward the end of 1938 two German chemists, Otto Hahn and Fritz Strassmann, found traces of barium in samples of uranium which they had been bombarding with neutrons. This result was elucidated the following year by Otto Frisch and Lise Meitner, who noted that the neutron bombardment split the uranium atom into two nearly equal parts in a process now known as fission. Hahn and Strassmann had produced the first artificial nuclear fission. The foundations for this discovery, however, had been laid during the previous two decades by scientists such as Bohr, Chadwick, Fermi, and Rutherford.[2]

The discovery of the nuclear fission reaction and the consequent release of vast amounts of energy, combined with the prospect of neutrons produced in the fission reaction inducing further fission reactions, gave rise to the possibility of achieving a fission chain reaction capable of releasing enormous quantities of energy. To some scientists this suggested the possibility that the nuclear fission reaction might make possible an explosive weapon capable of incredible destruction (although to many other scientists this possibility was rather remote). Nevertheless, in mid-1940, the U.S. government banned all publications on nuclear fission because of the potential for its military use.

As research continued through 1940 and 1941, it became apparent that nuclear weapons could be developed, and in early 1942 the program was turned over to the U.S. Army for military supervision. The task was directed from an office in the Manhattan district of New York and consequently came to be known as the "Manhattan Project."

A team led by Enrico Fermi was given the task of demonstrating the feasibility of a fission chain reaction by building a nuclear "reactor" using natural uranium and graphite. The world's first nuclear reactor was built in the squash courts under the stands of a football stadium at the University of Chicago.[3] It was constructed by piling up tens of thousands of graphite blocks, in some of which cavities had been made and filled with uranium oxide or metal, and hence became known as an "atomic pile." At 3.25 p.m. on December 2, 1942, a cadmium-plated control rod was withdrawn from the reactor and the first man-made, self-sustaining nuclear chain reaction was initiated.

The bulk of the world's known uranium reserves prior to the Second World War were contained in the very high-grade Shinkolobwe deposit in the Katanga province of the Belgian Congo (now Zaire). The Belgian company, Union Minière du Haut Katanga, became the first commercial producer of radium in the early 1920s, using material from the Shinkolobwe mine, and it had a virtual monopoly of the radium market until the mid-1930s. The uranium from the Shinkolobwe mine, however, was considered a worthless by-product of radium production and was left in waste heaps outside the factory in Belgium. In 1940 this "waste" fell into the hands of the Nazis when they invaded Belgium.

Union Minière's monopoly was broken in the mid-1930s when the Great Bear Lake deposit in Arctic Canada was developed. With competition, the radium market soon became saturated and the resulting price slump forced the closing of the Shinkolobwe mine.

At the outbreak of hostilities in 1939, an official of Union Minière, realizing the potential military use of uranium, arranged for ore from the Shinkolobwe mine, which had been stockpiled on the mine's closing, to be packed and shipped to the United States. The U.S. government subsequently purchased these shipments (which averaged an incredible 68 percent U_3O_8 as compared with 1 percent U_3O_8 from Great Bear Lake ore), and it eventually accounted for 80 percent of the total amount of uranium used in the Manhattan project.

While the successful construction of both a uranium and a plutonium atomic bomb in 1945 is well documented, throughout the war scientists had held out great hopes for the beneficial potential of the atom. Upon cessation of hostilities the United States Atomic Energy Commission was created with the primary goal of developing atomic power. Since the whole field of knowledge upon which reactor technology was based was strongly interlaced with the weapons program, there were many questions to be answered regarding reactor safety and regulation. It was not until 1957 that the first U.S. full-scale civilian nuclear power reactor went into operation at Shippingport (near Pittsburg), Pa. The world's first civilian nuclear-power reactor had been completed the previous year at Calder Hall in the United Kingdom. France (1958), the USSR (1958), and Canada (1962) followed closely behind. While many other countries built research reactors, it was not until the 1970s that extensive nuclear-power programs for generating

TABLE 1.1 Level of World[1] Installed Nuclear Capacity (GWe, year-end)

Country	1970	1975	1980
Canada	0.2	2.5	5.2
France	1.5	2.8	12.1
Germany (FRG)	0.8	3.3	8.6
Japan	1.3	5.9	14.3
Sweden	0.4	2.4	4.6
United Kingdom	5.3	5.3	6.6
United States	6.1	39.6	53.0
Others	2.4	6.7	10.9
total	18.0	68.5	115.3

[1] Excludes data for China, Eastern Europe, and the Soviet Union.

Source: Minerals Yearbook, U.S. Bureau of Mines, 1970 and 1975. *The Uranium Equation*, The Uranium Institute, Mining Journal Books, London, 1981.

electricity became a reality in most industrialized nations of the world. Table 1.1 illustrates the rapid growth in installed nuclear capacity which occurred in many Western nations between 1970 and 1980.

STRUCTURE OF THE BOOK

This book is concerned with the economics of uranium as determined by its commercial, as distinct from military, applications. Ignoring a few very minor applications, the sole use of uranium today is as an input in the nuclear fuel cycle. Thus the fortunes of the uranium industry are presently contained within the economic viability of nuclear power vis-à-vis its oil and coal-fired competitors and, ultimately, technological advances in the electricity generating industry.

The purpose of this book is to provide a thorough analysis of the economic structure of the uranium industry, leading eventually to the specification and estimation of an economic model of the U.S. uranium market. It will become apparent as the study progresses, that emphasis on the U.S. market is necessitated not only by the fact that the United States has dominated the world market with respect to both the demand and supply of uranium until recently, but also because of the paucity of data relating to the market elsewhere. The primary objective in developing such a model is to investigate the process of price formation in the short-term uranium market, with particular emphasis on the role of uranium inventories.

A number of studies of the U.S. uranium market have already been undertaken with the intention of producing forecasts of uranium supply, demand, and prices, but the majority of these studies have been based on engineering approaches to the problem.[4] A study by Ahmed[5] is of particular relevance to the work presented here, as he also developed an economic model of the U.S. uranium market and subsequently produced price forecasts for the period 1980–2000. While Ahmed's work provides a good (but now rather dated) introduction to the economics of nuclear power in the United States, his econometric analysis suffers from a number of major drawbacks. These will be considered in detail later in this study.

Since this book attempts to cover, in a fairly narrow sense, the economics of the uranium industry, a number of closely related issues stemming from the mining and use of uranium in the nuclear fuel cycle are afforded only scant treatment despite the fact that they may represent a major source of controversy. In particular, the environmental, social, and political controversies surrounding the various stages of the nuclear fuel cycle and the continuing debate over the association between nuclear power and the proliferation of nuclear weapons are relegated to the periphery of this study. This is in no way meant to reflect the author's opinion of their relative merits; rather it acknowledges that such debates constitute a separate study in their own right.[6]

A further omission is the impact on the nuclear-power industry, and hence the uranium industry, of long-term technological advances with not only alternative means of generating electricity (e.g., nuclear fussion) that may ultimately render nuclear fission redundant, but also with technological advances with reactors based on nuclear fission (e.g., the widespread introduction of breeder reactors).[7] Such factors represent very long-term prospects, advancing well into the next century and involve a time horizon well beyond the scope of this book. Currently the vast majority of the world's nuclear-power reactors are the U.S.-designed Light Water Reactors. Their dominance of the market is very unlikely to recede over the next 20 years, although they will undoubtedly become more efficient in their operations over this period of time. This study assumes that no radical technological advances such as those cited above will occur over the next 20 years. It seems to be a fairly safe assumption.

Finally, over a somewhat shorter time horizon, the uranium market will be influenced by the relative costs and benefits of using a nuclear-powered plant as compared with its oil-, gas-, and coal-fired counterparts or with hydroelectric power generation. Electricity utilities are the largest consumers of primary energy in the industrialized nations of the world, and represent the most capital-intensive sector of their respective economies. Thus any prospective investment in new generating capacity must be carefully evaluated on the basis of its technical merits, assurance of supply, political acceptability, and economic viability.

RELATIVE COSTS OF GENERATING ELECTRICITY

At the level of fuel costs prevailing in 1981, the Organisation for Economic Cooperation and Development (OECD) calculated that the total cost of nuclear-power electricity generation in member countries was significantly lower than that of electricity generation from oil-fired plants, while its competitive advantage over coal-fired plants varied from country to country.[8] Coal plants were more competitive in western parts of the United States and Canada due to relatively low-cost coal resources, while nuclear power had a cost advantage in the European Economic Community and Japan.

Economic comparison of electricity generation between nuclear and coal is difficult, however, because estimated costs (especially for nuclear power plants) are subject to large variations, both between and within countries. This is due to differences in reactor designs and regulatory requirements, in addition to the influence of lead times, interest rates, fuel prices, etc. Cost comparison studies, therefore, have to be treated with care.

Indicative cost estimates (for newly built plants) made by the OECD for representative oil-, nuclear-, and coal-based power generation in member countries are shown in Table 1.2. For oil, two cases are shown, one for a plant using low-sulfur, heavy fuel oil without flue gas desulfurization (FGD) and the other for high-sulfur fuel oil plant with FGD. In the case of coal, the costs of generation are estimated separately for the United States, Western Europe, and Japan, in order to compare the effects of differences in coal prices. Plants considered were assumed to operate at a 65

TABLE 1.2 Indicative Cost Estimates for Electricity Generation by Fuel (1981, U.S. mills/Kwh)

	Oil 2 × 600 MW		Nuclear	Coal with FGD 2 × 600 MW		
	Low Sulfur	High Sulfur with FGD	PWR 2 × 1100 MW	U.S.	Western Europe	Japan
Capital cost	10.8	12.9	24.8	17.1	17.1	17.8
Operating cost	2.5	4.2	4.2	5.1	5.1	5.1
Fuel cost	54.6	47.6	10.0	16.0	26.0	26.0
total cost	67.9	64.7	39.0	38.2	48.2	48.9
Reference						
Capital investment ($/Kwh)	577	692	1,331	920	920	956
(Initial investment $/Kwh)	(500)	(600)	(1,000)	(760)	(760)	(790)
(Interest during construction $/Kwh)	(77)	(92)	(331)	(160)	(160)	(160)
Construction lead times	3 years	3 years	6 years	4 years	4 years	4 years
Fuel cost	$33/bbl	$27/bbl		$40/t	$65/t	$65/t
($ per toe)	(238)	(194)	(40)	(60)	(100)	(100)
Conversion efficiency	37%	35%	34%	33%	33%	33%
Heat rate (K cal/Kwh)	2,300	2,450	2,500	2,600	2,600	2,600

Source: OECD International Energy Agency and Nuclear Energy Agency, *Nuclear Energy Prospects to 2000* OECD, Paris, 1982.

TABLE 1.3 Summary of Levelized Discounted Electricity Generation Costs (as of January 1, 1981)

Country	Ratio, Coal/Nuclear
Belgium	1.39
France	1.75
Germany (FRG)	1.64
Italy	1.57
Japan	1.51
Netherlands	1.29
Norway	1.42
Sweden	1.33
United Kingdom	1.43
United States	1.01

Source: OECD Nuclear Energy Agency, *The Costs of Generating Electricity in Nuclear and Coal-fired Power Stations,* OECD, Paris, 1983.

percent load factor for 30 years. An interest (and discount) rate of 10 percent was used.

Nuclear power clearly has a marked cost advantage over coal in Western Europe and Japan, while in the United States the indicative cost of the two modes is very similar. Oil presents a very high-cost alternative.

Country-by-country estimates of the relative cost of generating electricity by coal-fired and nuclear plants in the major industrialized nations of the OECD are given in Table 1.3. The ratio of the levelized electricity generation cost of coal to nuclear shows that coal-fired plants are considerably more costly than nuclear power throughout Europe, with the cost differential exceeding 50 percent in France, Federal Republic of Germany, Italy, and Japan.[9] Only in the United States are the two modes approximately equal in cost.

However, highly capital-intensive projects such as nuclear-power plants are particularly vulnerable to longer lead times and higher interest rates causing, respectively, an escalation of construction costs and higher capital charges. Table 1.4 shows the effects of a higher interest (and hence discount) rate (15 percent) and a longer lead time on the cost of nuclear-power generation. The effects of a higher interest rate on the cost of coal- and oil-fired plants are also shown in Table 1.4. With the interest rate raised to 15 percent, nuclear's cost advantage over coal is diminished in Western Europe and Japan, while in the United States coal shows a

TABLE 1.4 Effects on Generation Costs by High Interest and Longer Lead Time

	Oil		Nuclear		Coal		
Mills/Kwh	Low Sulfur	High Sulfur	Case A	Case B	U.S.	Western Europe	Japan
Capital cost	16.5	19.8	40.7	53.8	27.0	27.0	28.0
Operating cost	2.5	4.2	4.2	4.2	5.1	5.1	5.1
Fuel cost	54.6	47.6	10.0	10.0	16.0	26.0	26.0
total cost	74.1	71.6	54.9	68.0	48.1	58.1	59.1
Construction lead time	3 years	3 years	6 years	10 years	4 years	4 years	4 years
Interest rate	15%	15%	15%	15%	15%	15%	15%
Capital investment ($/Kw)	617	740	1,521	2,011	1,005	1,005	1,045

Source: OECD International Energy Agency and Nuclear Energy Agency, *Nuclear Energy Prospects to 2000,* OECD, Paris, 1982b.

significant cost advantage. When the higher interest rate is combined with a construction lead time of ten years, coal-fired plants represent a substantial cost advantage over nuclear in all three areas considered in the table. In fact, under these assumptions, the cost of electricity generated by nuclear power is only marginally below that of oil-fired plants. While economic factors are by no means the sole criterion for nuclear-power development, higher costs associated with long lead times could act as an additional disincentive.

At the time of writing this book, most industrialized nations were suffering from an oversupply of electricity generating capacity largely brought about by the combined effects of the industrial recession of the early 1980s and overly optimistic forecasts in the 1970s of electricity requirements for the 1980s. When significant additional capacity is required the relative costs of the various modes of electricity generation may well differ significantly from those presented in Tables 1.2 and 1.4. With unrequired capacity still coming on-stream or nearing the end of construction in many countries, however, decisions regarding extensive expansion of existing capacity are unlikely to be required for a number of years.

DEFINITIONS

Until the late-1970s, the United States dominated the world uranium market, accounting for the major share of both production and consumption. Thus most of the historical detail of the industry given in Chapter 3 refers to the American market. Over the past 30 years, government responsibility for matters concerning uranium has been entrusted with three different agencies. Until 1974, the controlling body was the United States Atomic Energy Commission (AEC); between 1974 and 1977, the Energy Research and Development Administration (ERDA); and, thereafter, the Department of Energy (DOE).

Although many centrally planned Eastern European countries and the Soviet Union have significant nuclear-power capacity and extensive supplies of uranium, data on their production levels and consumption requirements are kept confidential. With the exception of enrichment services provided by the USSR to a number of Western nations, and China's (rumored) small sales of uranium to "friendly" countries (believed to be Argentina, Japan, and

Yugoslavia),[10] there is very little East-West trade in any products associated with the nuclear fuel cycle. While China has plans to introduce nuclear power and possesses a complete fuel cycle capability as a result of its nuclear weapons program, again data on uranium resources and production are sparse. All references and data relating to "The World," therefore, refer to the world excluding China, Eastern Europe, and the USSR. Frequently this is referred to as the World Outside the Centrally planned economies Area (WOCA).

Throughout this study frequent reference will be made to reports published periodically since 1967 by the European Nuclear Energy Agency (ENEA) [which was renamed the Nuclear Energy Agency (NEA) when Japan became a full member on April 20, 1972] of the Organisation for Economic Cooperation and Development (OECD), in conjunction with the International Atomic Energy Agency (IAEA). These reports, commonly referred to as the "Red Book," contain estimates of WOCA uranium resources, production, and demand. To date, ten such reports have been published:

(1) "World Uranium and Thorium Resources," OECD (ENEA) Report, Paris, 1965.

(2) "Uranium Resources, Revised Estimates," Joint OECD(ENEA)/IAEA Report, Paris, 1967.

(3) "Uranium Production and Short Term Demand," Joint OECD(ENEA)/IAEA Report, Paris, 1969.

(4) "Uranium Resources, Production and Demand," Joint OECD(ENEA)/IAEA Report, Paris, 1970.

(5 to 10) "Uranium Resources, Production and Demand," Joint OECD(NEA)/IAEA Reports, Paris, 1973, 1975, 1977, 1979, 1982, 1983.

A recent (1982), more detailed analysis of the uranium demand situation is contained in the report by the Working Party on Nuclear Fuel Cycle Requirements (WPNFCR) "Nuclear Energy and Its Fuel Cycle: Prospects to 2025," published by the Nuclear Energy Agency of the OECD and commonly known as the "Yellow Book." A previous Yellow Book appeared in 1978 and, in the interim period, similar material was published by the International Nuclear Fuel Cycle Evaluation (INFCE) Working Group 1 (WG1) of the IAEA in 1980.

UNITS OF MEASUREMENT

The rate of using or producing energy is generally measured by a unit of power called a kilowatt. One thousand kilowatts are equal to a megawatt (MW). MWe denotes the electrical power output of a power station. This should not be confused with the thermal power of a power station, which measures the rate at which heat is produced (by fission in the reactor core if it is a nuclear power station). Typically the MWe is only about one-third of the thermal power for a Light Water Reactor and up to 40 percent for a modern fossil-fuel-fired power station. This ratio is called the thermal efficiency of the power station. In this study nuclear capacity is frequently measured in gigawatts electrical (GWe), one GWe being equivalent to 1,000 MWe.

The price of uranium is conventionally expressed in U.S. dollars per pound of uranium oxide (*$/lb* U_3O_8). In recent years, however, a number of countries have moved to the metric measurement $/kg U_3O_8, while the OECD/IAEA reports express prices in terms of kilograms of uranium metal ($/kg U), where (approximately) $1/lb U_3O_8 = $2.2/kg U_3O_8 = $2.6/kg U. Price is usually quoted on a "delivered" basis; i.e., it includes items such as transport, packing, and insurance costs. In the ensuing text and tables, all prices are quoted in *$/lb* U_3O_8, and all dollar signs denote U.S. dollars unless specified otherwise.

All quantity data are reported in short tons (i.e., 2,000 lbs) U_3O_8. The imperial (or "long") ton is equivalent to 2,240 lbs, while the metric ton (or tonne) is equivalent to 2,205 lbs. The approximate conversion factors are:

$$1 \text{ tonne U} = 1.30 \text{ short tons } U_3O_8 = 1.18 \text{ tonnes } U_3O_8$$

In the text all quantity data will be expressed in (short) tons U_3O_8.

NOTES

1. Klaproth was one of the leading chemists of the era. He is also credited with the discovery of beryllium, tellurium, chrome, cerium, and zirconium.
2. This period in the story of uranium has been documented in detail by Bickel (1979).

3. The story of the nuclear age since the discovery of fission has been traced by Goldschmidt (1982).
4. Charles River Associates (1981) provide a synopsis of eight major studies of price projections for the U.S. uranium market (using six different approaches) and discuss the relative merits of the various methodologies used in these studies.
5. Ahmed (1979).
6. For a discussion of these issues in a general context, the reader is referred to the Ranger Uranium Environmental Inquiry, First Report (1976) or The Flowers Report (1976). The more local environmental aspects which relate particularly to the Ranger proposal appear in the Ranger Inquiry's Second Report (1977).
7. Amr et al. (1981) provide a discussion of the feasibility of alternative methods for generating electricity which may become a reality in the early years of the next century.
8. OECD International Energy Agency and Nuclear Energy Agency (1982). Another important study is by the U.S. Department of Energy (1982) on the projected costs of electricity from nuclear- and coal-fired plants in 1995. Also see Amr et al. (1981).
9. The "levelized" cost is the average cost in constant money terms per unit of electricity fed into the grid at which the total lifetime output of the plant exactly balances the costs of the plant, its operating and fuel costs, plus waste management and ultimate decommissioning.
10. Reported by Geddes (1983).

2

The Nuclear Fuel Cycle

INTRODUCTION

A nuclear-power plant is simply a very large and expensive device for generating electricity, using the heat generated by nuclear fission, a physical reaction. This differs from coal-, or gas-fired power plants, which generate heat by combustion, a chemical reaction. Once the heat has been generated, however, the remaining steps in the process of electricity generation are very similar. The heat is used to produce steam, which drives turbines that turn electric generators.

Uranium is a totally different fuel from its fossil fuel competitors. The latter are carbon compounds which react with oxygen when they burn, producing heat and carbon dioxide. The chemical elements present before combustion are still present in the same quantities afterwards. The production of energy by fission, however, involves the conversion of a small part of the original mass into energy. Because of this, the combined mass of the products of fission is slightly less than the mass of the original uranium.

This chapter presents a simple description of the fission process, current reactor types, and nuclear fuel cycles. For a more detailed description of the economic and technical aspects of power reactor types and nuclear fuel cycles, the reader is referred to Duderstadt (1979) or IAEA (1982a).

NUCLEAR FISSION

An atom is the smallest particle into which an element can be divided chemically. It consists of a nucleus of positively charged particles called protons and a number of uncharged (or neutral) particles called neutrons. The positive charges on the protons are balanced by an equal number of negatively charged electrons in motion around the nucleus. The number of protons in the nucleus determines the chemical properties of an element. Uranium has 92 protons. Different atoms of the same element can have different numbers of neutrons; such atoms are known as isotopes of the element. For example, uranium-238 has 146 neutrons, whereas uranium-235 has only 143 neutrons. Both contain 92 protons. Thus their chemical properties are identical but their nuclear properties are very different.

Nuclear fission refers to the process brought about by the splitting (i.e., "fission") of certain heavy elements by neutron bombardment. Uranium is the only "fissionable" element which occurs in large quantities in nature. However, only one natural isotope of uranium is fissionable—uranium-235—and this makes up only 0.711 percent of naturally occurring uranium. The nonfissionable isotopes are uranium-234 (0.006 percent) and uranium-238 (99.283 percent). While uranium-234 is too scarce to be of commercial use, uranium-238, along with another natural element thorium-232, is called "fertile" because it can be converted into fissionable material when placed inside a nuclear reactor.

When the nucleus of a fissionable atom absorbs a neutron, it has a high probability of splitting, or fissioning, into two smaller nuclei with the consequent release of a significant amount of energy and the emission of two of three neutrons. The emitted neutrons are available for absorption by other fissionable nuclei. A fission reaction can be made self-sustaining if the neutrons emitted by one fission reaction go on to trigger a subsequent fission reaction. Such a process generates great quantities of energy which are utilized in power reactors, directly or indirectly, to make steam, which can then be used to generate electricity. Not all of the neutrons emitted in a fission reaction go on to initiate subsequent fissionings; some escape from the region of the reaction, some are absorbed, and some are captured by heavy nuclei without occasioning fissioning. The rate of such fissions must be controlled, however, so that one fission produces (on average) just one new fission. If new fissions were to be set off by more than (an average of) one of the emitted neutrons, the rate of fissions and heat output would increase exponentially.

In general, a slow neutron is more likely to cause fission than a fast one as it is more likely to interact with a fissionable atom before escaping from the reactor core. Uranium-235 at low concentrations requires a moderator to slow down neutrons to ensure sufficient fission to sustain the chain reaction. Ordinary water (H_2O) serves the purpose if the concentration of uranium-235 in the uranium is more than 1 percent. U.S. reactors typically use a concentration of around 3 percent uranium-235. The process by which the percentage of fissionable atoms in the uranium is increased from 0.7 percent to 3.0 percent is known as uranium enrichment. Alternatively, heavy water (D_2O) or graphite can be used as moderators. If heavy water is used, a chain reaction can be sustained using natural uranium with a concentration of only 0.7 percent uranium-235. To make heavy water, oxygen is combined with deuterium, an isotope of hydrogen that contains one extra neutron. Hence heavy water weighs slightly more per molecule than does ordinary (usually referred to as "light") water. Most U.S. reactors are moderated by light water, whereas the CANDU (Canadian deuterium uranium) reactors developed by the Canadian government use heavy water. Graphite is used as a moderator in some Western European-constructed reactors, but as we will see, these are of relatively minor importance.

We have already noted that uranium-238, while not fissionable itself, is, together with thorium-232, a source of fissionable material. If a neutron strikes a uranium-238 nucleus fission is most unlikely. Neutron absorption causes a series of nuclear transmutations in uranium-238. The end result is plutonium-239, an element that is fissionable. Similarly, thorium-232 will convert upon neutron capture into uranium-233, also a fissionable material. These two neutron-capturing isotopes are said to be "fertile."

NUCLEAR-REACTOR COMPONENTS

While there are many different types of fission reactors, six basic components are common to the majority of them: fuel, moderator, coolant, control elements, reflector, and vessel.

Fuel. The core of the reactor contains the uranium fuel and this is where the heat is generated. The majority of reactors in operation today (the American-designed Light Water Reactors) use enriched uranium as fuel, so that it contains about 3 percent of the fissile isotope uranium-235, rather than the 0.7 percent which occurs naturally. The CANDU reactors use natural uranium. Uranium-233 and plutonium-239 (produced by conversion of thorium-232 and uranium-238, respectively) can also be used as fuel, generally in combination with uranium-235.

Moderator. Nuclear reactors fall into two broad categories: "thermal" and "fast." Thermal reactors maintain the chain reaction with slow neutrons that have lost most of their original energy by collision with nuclei of a moderator. If the uranium has been enriched to 3 percent uranium-235, then light water can be used as a moderator. Heavy water can be used if the fuel is natural uranium. Fast reactors use uranium highly enriched in uranium-235, or high concentrations of other fissile material, as fuel. Because the concentration of fissile nuclei is so high, the chain reaction can be sustained by fast neutrons and no moderator is needed.

Coolant. The heat produced by fission is carried away by liquid or gaseous coolants circulating through the core. Water is the most common coolant for thermal reactors, although heavy water, liquid metals (especially sodium), carbon dioxide, and helium are also used. In the majority of reactors, the hot coolant passes through a heat exchanger in which water in another circuit boils to produce steam for the turbine. Alternatively, the coolant may boil and the steam may be used directly to drive the turbine.

Control Elements. The rate of fission in a reactor is controlled by inserting or withdrawing neutron-absorbing rods into the core. The deeper these control rods are inserted in the reactor, the more neutrons they absorb. Partial withdrawal of the rods allows enough neutrons to circulate to start the chain reaction. Full insertion stops the reaction.

Reflector. The core of a reactor is surrounded by a reflector, the purpose of which is to reflect back into the core some of the neutrons which might otherwise escape. For thermal reactors, good moderating materials also make good reflectors. Fast reactor reflectors are usually made of fertile material.

Vessel. The nuclear part of the reactor is contained in a reactor vessel. If the coolant is maintained at a high pressure, the reactor vessel is referred to as a pressure vessel.

NUCLEAR-POWER REACTOR SYSTEMS

Reactors are commonly classified as burners, converters, and breeders, depending upon the amount of new fissile material formed (from the fertile material) during reactor operations. The conversion ratio is the ratio of the number of fissile nuclei formed to the number consumed. A reactor in which there is little conversion (i.e., it has a low conversion ratio) is termed a burner; one in which the conversion ratio approaches 1 is called a converter; and any reactor in which the conversion ratio exceeds 1 is called a breeder.

The reactors which are most common in commercial operation or currently stressed in reactor development programs are described below.

Light Water Reactor (LWR)

In LWRs light (or ordinary) water acts as both the moderator and the coolant. As a consequence, the uranium which is used as fuel must be enriched to increase the concentration of U-235, thus necessitating the availability of uranium enrichment facilities.

There are two kinds of LWR: the pressurized-water reactor (PWR) and the boiling water reactor (BWR). Schematic diagrams of typical examples of these types of power reactor are shown in Figure 2.1. In the BWR the coolant is allowed to boil until it becomes high-pressure steam and thus can be used directly to drive turbine generators, whereas with the PWR the heat is transferred by the coolant to a secondary system for the generation of steam.

At year-end 1983, there were 238 nuclear reactors (of at least 150 MWe gross) in operation in Western nations with a gross installed capacity of 174.8 GWe. From Table 2.1 it can be seen that LWRs accounted for 86 percent of total capacity, this figure being split between PWRs and BWRs in a ratio of about 2:1.

Heavy Water Reactor (HWR)

Since heavy water is a more effective moderator than light water, the HWR uses natural, unenriched uranium as a fuel. The

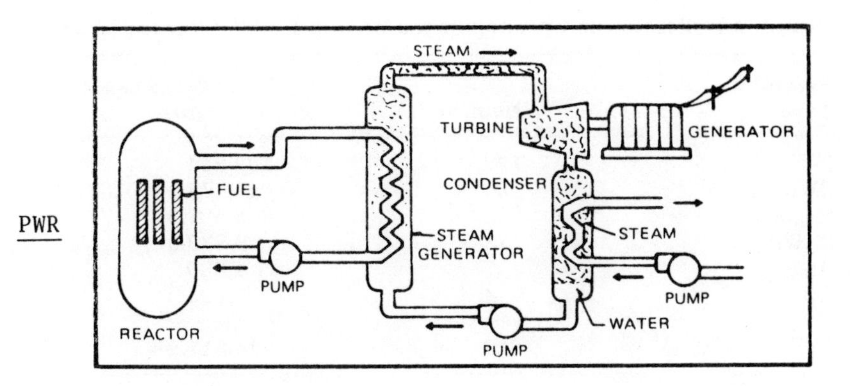

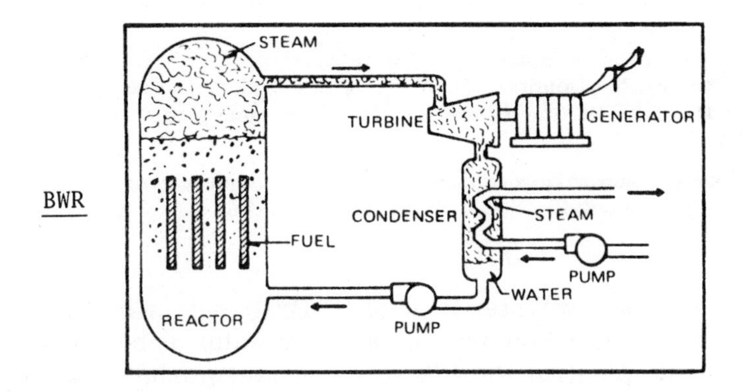

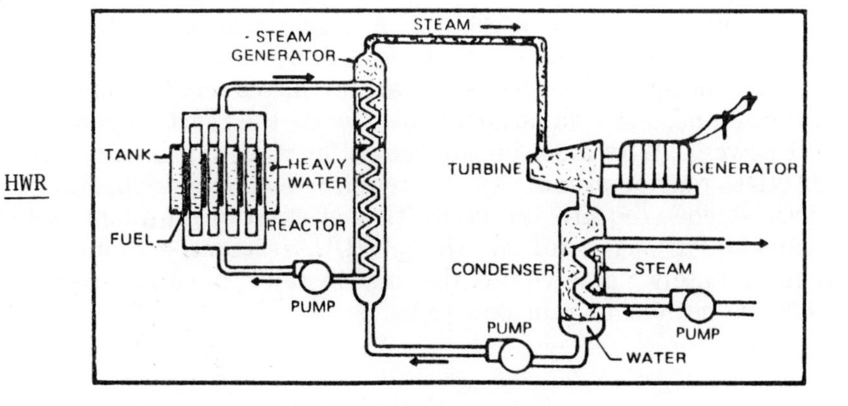

FIGURE 2.1 Schematic Diagrams of Typical Power Reactors.
Source: International Atomic Energy Agency. *Guidebook on the Introduction of Nuclear Power,* IAEA, Vienna, 1982a.

TABLE 2.1 Reactors in Operation in the Western World[1] (year end 1983)

Reactor Type[2]	Number	Total design capacity, MWe (gross)
PWR	122	102,814
BWR	61	47,178
Magnox	26	8,527.3
PHWR	18	9,973.9
AGR	7	4,600
Other[3]	4	1,724
total	238	174,817.2

[1] Operation is defined as first electricity generation (not commercial operation, which can be delayed for varying periods). Only reactors of 150 MWe gross and above are included.

[2] See text for an explanation of the abbreviations.

[3] Two FBRs [Dounreay (U.K.)—270 MWe; Phenix (France)—250 MWe], one HTGR [Fort St. Vrain (U.S.A.)—342 MWe], one graphite/water [Hanford (U.S.A.)—862 MWe].

Source: Laurie Howles, "Nuclear Station Achievement," *Nuclear Engineering International,* 29, No. 355 (May 1984): 36–38.

Pressurized Heavy Water Reactor (PHWR), of which the CANDU reactor is the most common, uses heavy water as both moderator and coolant. As with the PWR, the coolant produces steam in a second circuit (see Figure 2.1). The Steam-Generating Heavy Water Reactor (SGHWR), however, uses heavy water as a moderator and light water as a coolant. In common with the BWR, the coolant boils to provide steam direct for the turbine.

The design of the CANDU reactor enables fuel elements to be inserted and removed automatically while the reactor operates at full power. LWRs must be shut down for refueling. Although the CANDU reactor eliminates the requirement for enrichment services, it does require the production of substantial quantities of heavy water. A total of 18 CANDU reactors accounted for approximately 5.5 percent of the total installed nuclear capacity of Western nations at year-end 1983.

Gas-cooled Reactor

Gas- (pressurized carbon dioxide) cooled Magnox reactors using natural uranium fuel rods encased in "magnox," a magnesium alloy, and a graphite-moderated core have been in use for many years in the United Kingdom and represent an established technology.

More recent design improvements have led to the advanced gas-cooled reactor (AGR), which employs enriched uranium as fuel (about 2 percent uranium-235), which is in the form of uranium dioxide. This allows a higher proportion of the uranium-235 to be consumed before the fuel rods are replaced and enables the reactor to operate at higher temperatures, giving greater thermal efficiency.

Although the Magnox and AGR reactors (combined) accounted for nearly 14 percent of the total number of nuclear-power reactors in the Western world at the end of 1983, their combined capacity was only about 7.5 percent of the total. Both these figures are expected to decrease rapidly as LWR-based nuclear-power programs continue to expand on a worldwide basis.

High-temperature Gas-cooled Reactor (HTGR)

The HTGR makes use of three fuels—highly enriched uranium (over 90 percent uranium-235), uranium-233, and thorium-232 (for conversion to uranium-233)—with the mix of fuels changing over the reactor lifetime. The initial loading consists of highly enriched uranium and thorium-232; in subsequent loadings recycled uranium-233 would replace highly enriched uranium. Graphite is used as a moderator. Thermal efficiency is increased because the coolant gas, helium, is allowed to reach much higher temperatures than other forms of coolant.

Because of the slower than anticipated rate of growth of the electric power industry during the late 1970s and early 1980s, all but one of the HTGRs that were on order in the United States have been canceled and the HTGR has yet to demonstrate economic feasibility. The sole exception, a 342-MWe HTGR at Fort St. Vrain in Colorado, has been operating since 1979 but not without serious initial commissioning and operating problems.

Liquid Metal Fast Breeder Reactor (LMFBR)

All of the reactors described above are thermal reactors. "Fast" reactors have been in the developmental/pilot plant stages for many years, although their progress has been severely impeded not only by technological and economic constraints but by the political constraint of the specter of a "plutonium economy." While the United States was an early leader in breeder reactor development, the past decade has witnessed a substantial reduction in the nation's development effort. Some Western European nations, however, have continued their programs, and demonstration "fast" reactor plants in the 200- to 300-MWe class are currently operating in France, the United Kingdom, and the USSR. The Federal Republic of Germany's first demonstration breeder plant is scheduled to begin operations in 1987. The world's first commercial-sized LMFBR (i.e., 1,200 MWe) plant at Creys Malville in France is scheduled to attain initial criticality in early 1985.

The LMFBR fuel rods contain a mixture of about 20 percent plutonium oxide with depleted uranium dioxide, and a blanket of rods containing depleted uranium dioxide surrounds the core. Neutrons released by the reactor fuel convert fertile material (e.g., uranium-238), located in the blanket, to fissile material (plutonium-239) faster than the fissile material in the fuel is converted to nonfissile material. Hence the term "breeder" reactor. The initial loading could use either plutonium recovered from spent LWR fuel or enriched uranium; subsequent loadings would employ plutonium bred in the LMFBR itself. Liquid sodium is used as a coolant.

Figure 2.2 summarizes the pertinent features of the different forms of nuclear-power reactors outlined above.

THE NUCLEAR FUEL CYCLE

The manufacture of fuel for nuclear-power stations and its processing and management subsequent to reactor discharge are sometimes referred to as the "front end" and "back end" of the nuclear fuel cycle. In between lies the irradiation—i.e., the period when the fuel is contained within the core of the reactor.

The total fuel cycle comprises a number of activities, the possible combinations of which provide the various fuel cycle options. These activities are:

- uranium mining and milling
- uranium refining and conversion to hexafluoride
- uranium enrichment
- fuel fabrication

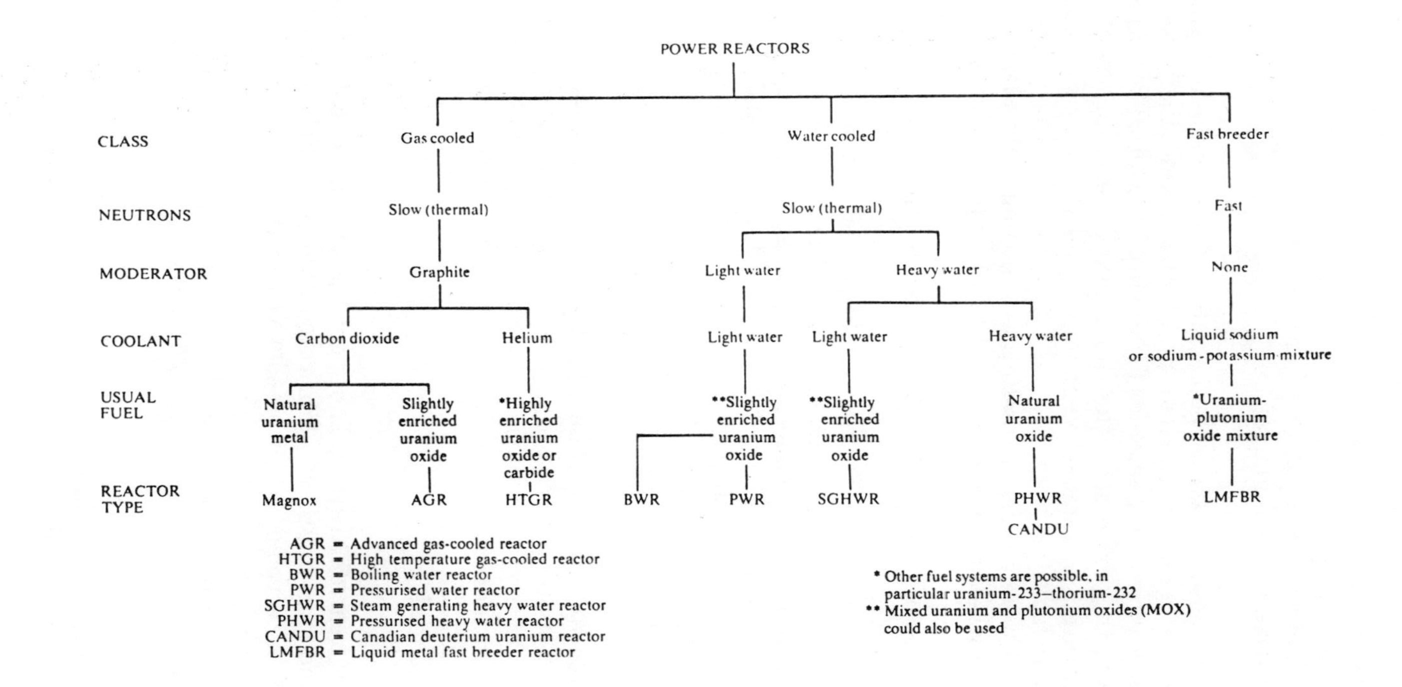

FIGURE 2.2 Reactor Types.
Source: Ranger Uranium Environmental Inquiry, First Report, Australian Government Publishing Service, Canberra, 1976.

- reactor operation
- spent fuel storage
- spent fuel reprocessing
- radioactive waste management

Each of these activities potentially affects human health, safety, the environment, nuclear proliferation, and theft. Each activity has economic implications for the final product: electricity. In addition to the above activities, the fuel cycle also incorporates the transportation of radioactive material and the decommissioning of nuclear facilities.

Three different types of fuel cycle are commonly identified, depending on whether or not the spent fuel is reprocessed and, if it is, to what type of reactor the uranium and plutonium are recycled. The three types of fuel cycle are illustrated in Figure 2.3.

Once-through Fuel Cycle. Strictly speaking this is not a cycle since the unused part of the spent fuel is not recycled. The spent fuel is not reprocessed but is kept in temporary storage until it can be sent for permanent disposal by, for example, conditioning it and placing it underground in a deep geological repository. The vast majority of nuclear-power plants in operation today have once-through fuel cycles.

Thermal Reactor Recycle. The spent fuel is reprocessed. The uranium and plutonium are separated from the fission products which are sent for permanent disposal. The uranium and the plutonium can then be fabricated into new fuel elements and thus recycled to the same type of reactor from which the plutonium was initially produced.

Fast Breeder Reactor Recycle. The spent fuel is reprocessed and the uranium and plutonium are fabricated into new fuel elements. They are then recycled to fast breeder reactors.

Choice of the appropriate fuel cycle depends on a vast interdependent set of variables. In general terms they are:

1. Economics of recycling,
2. Availability of uranium,
3. Availability of other sources of electrical energy,
4. Environmental, political, safety, and technological factors.

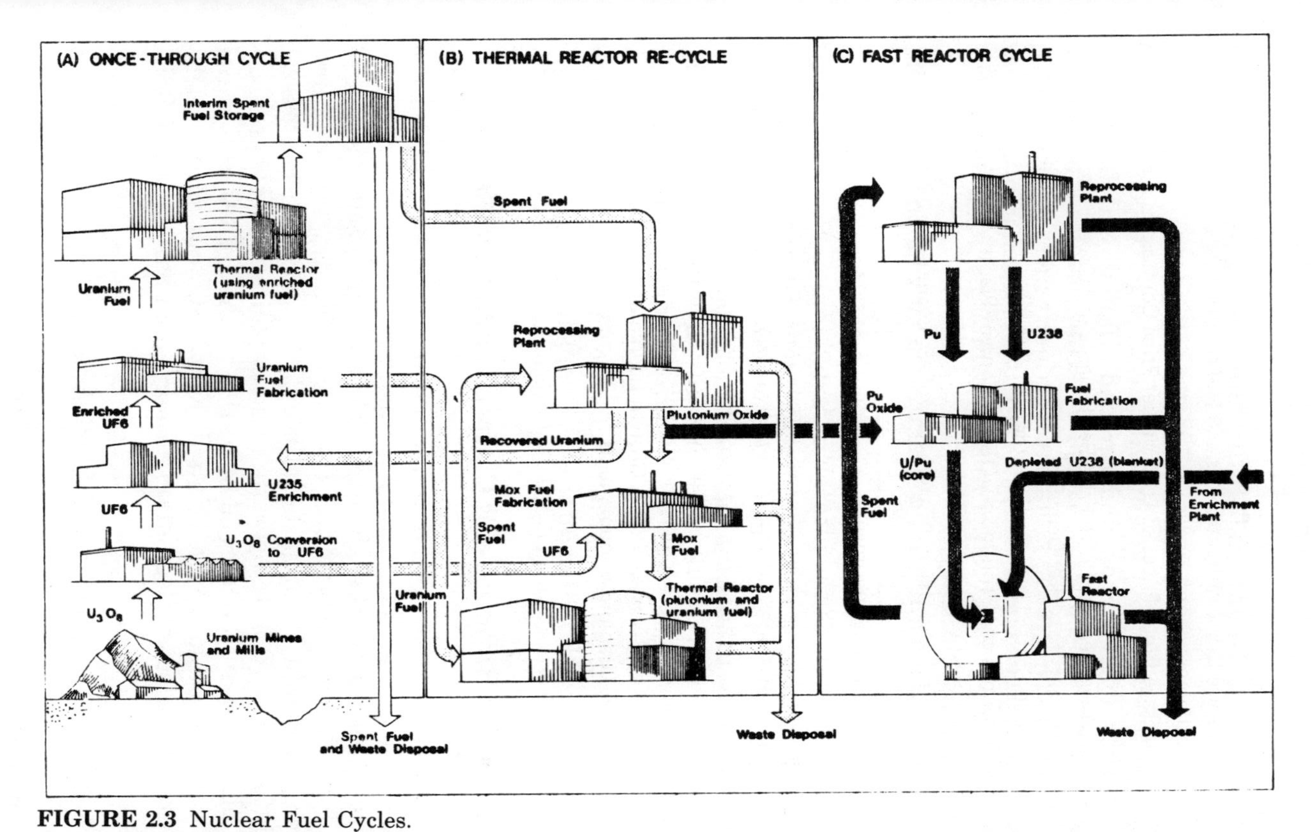

FIGURE 2.3 Nuclear Fuel Cycles.
Source: International Atomic Energy Agency. *Guidebook on the Introduction of Nuclear Power,* IAEA, Vienna, 1982a.

STAGES OF THE FUEL CYCLE

Uranium Mining and Milling

Uranium ore is mined using techniques and machinery which are similar to those used in mining other ores, e.g., copper, lead, and zinc. Either open-cut or underground mining will be chosen, depending on a number of complex factors, such as ore depth, deposit size, ore grade, ground conditions, surface topography, environmental considerations, etc. To date, solution mining (*in situ* or heap leaching) has been of relatively minor significance in the uranium mining industry.

Since even a significant concentration may mean that the ore contains less than 0.1 percent uranium, milling and preconcentration are necessary to minimize transport and chemical extraction costs. The uranium is chemically extracted from the ore (or milled) to produce a concentrate containing about 80 percent uranium oxide (U_3O_8), which is commonly known as "yellowcake." Yellowcake is the form in which uranium is most commonly traded.

Since the ore typically contains only about 0.1 percent U_3O_8, the rest is rejected as tailings. The tailings constitute a potential health hazard since they release radioactive radon gas to the atmosphere. To overcome this problem, they are stored in a tailings dam and ultimately, after mining is complete, buried.

Uranium Refining and Conversion to Hexafluoride

The yellowcake is shipped to refining plants where it is purified to produce nuclear-grade uranium compounds. The majority of nuclear reactors require that the uranium be enriched before it is used as a fuel. Enrichment requires that the uranium concentrate first be converted into uranium hexafluoride (UF_6), which is a gas at conditions near room temperature and pressure. Two UF_6 conversion plants are in commercial operation in the United States and one each in Canada, the United Kingdom, and France. UF_6 conversion is also carried out on a commercial basis by the USSR.

Uranium Enrichment

Natural uranium contains only 0.7 percent of the energy-producing isotope, uranium-235. The remainder of the natural uranium, uranium-238 (ignoring the minute amount of uranium-234), is the nonfissile part. Uranium enrichment is the process by which natural uranium is physically (as distinct from chemically) altered into a richer mixture of the fissile isotope uranium-235 to a level of about 3 percent, the balance being uranium-238. The remaining uranium is depleted to a level which is determined by technical and economic balances.

Uranium enrichment services are sold in Separative Work Units (SWUs), which is a measure of the amount of effort required to separate uranium-235 from uranium-238. The proportion of uranium-235 remaining in the depleted uranium (the tails) after enrichment is called the tails assay. It is typically between 0.2 and 0.3 percent. The higher the tails assay, the less will be the amount of enrichment energy (SWUs) required to produce a given quantity of enriched uranium, but the larger will be the amount of natural uranium needed to produce that quantity of the desired product.

Uranium enriched to very high levels can (potentially) be used to construct nuclear weapons. Enrichment technology, therefore, is classified and all enrichment agencies are government-controlled. At present there are four enrichment agencies in the world; the United States, the Soviet Union, and two consortia of Western European nations operate one each. Pilot enrichment plants are either operational, or under construction, in Brazil, Japan, and South Africa.

Four methods of enriching uranium are of current interest: gaseous diffusion, gas centrifuge, aerodynamic processes, and laser processes. Gaseous diffusion is the established technology, having been in large-scale operation for more than 25 years in the United States, while France and the United Kingdom have been operating small plants for a considerable time. The gas centrifuge process, a relatively recent addition to the commercial enrichment market, has two major advantages over gaseous diffusion: it is more flexible in matching capacity with demand and it is more energy efficient. Both South Africa and the Federal Republic of Germany have been developing aerodynamic processes, and the latter has sold its technology to Brazil. Laser enrichment is still largely experimental.

A more detailed analysis of the economics of uranium enrichment is reserved for Chapter 10. We continue with a detailed description of the alternative technologies for uranium enrichment which were outlined above.

Gaseous Diffusion

Uranium hexafluoride (gas) under pressure is filtered (diffused) through a porous membrane. A slightly higher percentage of the lighter U-235 isotopes diffuse through the barrier than U-238 isotopes, thus producing a more enriched mixture on the far side of the barrier and leaving a slightly depleted mixture on the original side. Since a single stage of this diffusion process increases the concentration of U-235 isotopes by a factor of only 1.00429, the uranium hexafluoride must go through a cascade of more than a thousand such stages of diffusion to enrich the gas from the original 0.7 percent to the higher concentration (usually about 3 percent) needed for reactor fuel. The depleted product is known as the "tails."

The gaseous diffusion stages require large amounts of power. Thus even the smallest practical gaseous diffusion plant is a huge establishment costing several billion dollars and requiring the equivalent of two or three nuclear-power plants to operate it. For most countries this would be an uneconomic proposition. Three gaseous diffusion enrichment plants are currently operating in the United States, at Oak Ridge (Tennessee), Paducah (Kentucky), and Portsmouth (Ohio), and a consortium of Western European nations (France, Belgium, Italy, and Spain), known as Eurodif, has a gaseous diffusion enrichment plant at Tricastan in France.

Gas Centrifuge

Gas centrifuge is the only other enrichment technology that has reached commercial use. When the uranium hexafluoride enters a spinning centrifuge, the heavier molecules tend to concentrate to the outside surface of the centrifuge, leaving the lighter molecules containing uranium-235 nearer the center. The gas is thus divided into two streams, one slightly enriched, and the other slightly depleted. Centrifuge stages are characterized by much larger separation factors than the gaseous diffusion stages, ranging from 1.1 to 1.4. Thus only a relatively few centrifuges must be connected in series to achieve substantial enrichment. The flow rates possible in centrifuges, however, are much lower than in gaseous diffusion stages and consequently large numbers of centrifuges in parallel are required for appreciable enrichment capability.

The appeal of gas centrifuge technology is that the minimum economic size of enrichment plant is much smaller than for gaseous diffusion, and the former can be expanded in stages. In addition, its energy requirements are only 10–15 percent of those required by comparable gaseous diffusion plants.

Two gas centrifuge enrichment plants (operated by URENCO, a British, Dutch, and FRG organization) are currently in commercial operation at Capenhurst (United Kingdom) and Almelo (Netherlands). A planned addition to U.S. enrichment capacity involves the construction of a gas centrifuge plant at Portsmouth, which is scheduled for completion around 1990. Japan is currently operating a pilot gas centrifuge plant at Ningyo-Toge.

Closely related to the centrifuge is the Becker nozzle separation method. In this process, jets of uranium hexafluoride, mixed with hydrogen, are pumped at high speed through curved tubes. The lighter molecules containing uranium-235 tend to move off at slightly greater angles than the heavier ones, and the enriched and depleted streams are separated at the nozzle. As in the other processes, the gas has to pass through many stages before it reaches the required level of enrichment. The advantage of this process is that it requires only modest technology compared to centrifuge enrichment. The major drawback, however, is that it is even more energy inefficient than gaseous diffusion. A demonstration enrichment plant using the (FRG) nozzle enrichment process is currently being constructed in Brazil.

South Africa is developing a pilot enrichment plant based upon the "helikon" enrichment process, which is closely related to the centrifuge and nozzle methods. It also requires considerable amounts of energy.

Laser Isotope Separation

It is possible to utilize laser technology for uranium enrichment. Laser separation is based on exploiting the slight differences in excitation energies of U-235 and U-238 atoms (evaporated from metal) or molecules (UF_6). The required isotope, having been selectively ionized, is then separated from the neutral atoms through electric or magnetic fields. The alternative way consists of inducing photochemical reactions of selectively excited UF_6 molecules. The reaction products are then physically or chemically separated from the UF_6.

The potential advantages of laser technology for isotope separation include large separation factors (as large as 10), modest power requirements, and (probably) significantly lower separation costs. The principal disadvantages are the significant technical problems which must be overcome before isotope separation can be applied on a commercial scale.

Fuel Fabrication

The final step in the front end of the fuel cycle takes place when the enriched uranium (or natural UF_4 for HWR fuel) is converted into uranium dioxide (UO_2), formed into small ceramic pellets, and encapsulated in a thin tube made of a suitable cladding material, chiefly zircaloy, to form the fuel rods. The fuel rods are then assembled into bundles, called "fuel assemblies." The number and arrangement of fuel rods in the fuel assemblies are determined by the specifications of the reactor core design. Mixed oxide fuels would be required for thermal recycle programs and FBRs.

The fabrication of fuel elements is a well-established technology and, unlike the enrichment process, fuel element fabrication plants operate on a commercial basis. From the point of view of safety and reactor performance, the fuel fabrication is a critical step in the fuel cycle and extensive tests and inspections are carried out during all parts of the operation.

Reactor Operation

The fuel assembly is inserted into the power reactor where the fuel is used to generate heat through fission. From this point on, the system follows the conventional steps of using the heat to produce steam, which in turn drives steam turbines that turn electric generators.

The initial fuel loading of a typical 1,000-MWe LWR power plant contains about 80 tonnes of slightly enriched (about 3 percent) uranium-235. These fuel elements must be removed from the reactor before all the fissile material (uranium-235 and plutonium produced by conversion of uranium-238) is consumed. The buildup of fission products also constitutes parasitic absorbers of the neutrons and threatens the mechanical integrity of the fuel elements. A LWR is shut down annually for a period of several weeks, during which about one-third of the fuel elements are replaced with fresh ones. CANDU reactors are refueled continuously without interrupting their operation.

Spent Fuel Storage

Upon its discharge from the reactor, the spent fuel contains unconsumed uranium, fission products, plutonium, and some other heavy elements.[1] It generates heat and is radioactive and must be stored underwater at the plant site in specially constructed ponds to remove the heat generated by radioactive decay. Afterward, depending on the fuel cycle strategy adopted for ultimate disposal of the spent fuel or reprocessing, the fuel will be removed from the cooling ponds to be sent for permanent disposal or to a reprocessing facility.

Spent Fuel Reprocessing

Following a cooling-down period of about one year in the temporary storage pond, during which the most intense short-lived and intermediate half-life radioactive fission products have decayed, spent fuel can be transferred to a reprocessing plant to separate residual uranium and plutonium from the fission products. The recovered uranium, which might still contain around 1 percent U-235, can be converted into uranium hexafluoride for subsequent reenrichment (i.e., it is recycled). The recovered plutonium can be converted into plutonium dioxide for subsequent use in mixed oxide fuel elements (i.e., a blend of uranium and plutonium dioxides). The fission products constituting the radioactive waste have to be treated and disposed of.

Uranium enrichment and reprocessing of spent fuel are considered the most sensitive elements of the fuel cycle from the nonproliferation point of view. Fuel reprocessing is the stage at which plutonium is recovered. While this plutonium is not "weapons grade," nevertheless, given scientific expertise and a lot of luck, it could be made into a device that might explode. Transfer of technology and international cooperation in this field have therefore been very limited and subject to tight restrictions.

A number of countries have laboratory, pilot plant, or full-scale reprocessing facilities in operation or planned. Currently, only France and the United Kingdom have industrial-scale reprocessing plants in operation (for oxide fuel from LWRs in France and metallic fuel from Magnox reactors in the United Kingdom), but five more countries have plans to introduce such plants. At present, only a small amount of spent fuel is being reprocessed, and even by 1990 when some large-scale commercial plants could be operational, less than half of the required reprocessing capacity will be available. Delays in reprocessing, however, are connected not only with technical aspects of construction of reprocessing facilities or difficulties in FBR developments, but also with the political and institutional problems of nonproliferation.

Radioactive Waste Management

The final steps in the nuclear fuel cycle are the management and disposal of the radioactive wastes, some of which may have radioactive half-lives extending to tens of thousands of years.[2] This is the most controversial aspect of nuclear power, and it is probably in this sector of the fuel cycle that most development is to be expected in the coming years.

Any material in any form containing or contaminated with radioactive nuclides which cannot be safely released to the environment and for which there is no future intended use is defined as "radioactive waste." Some form of waste is generated at nearly all stages of the nuclear fuel cycle, ranging from the tailings at the milling stage to the high-level waste from fuel reprocessing operations. Whenever practicable, radioactive waste materials are concentrated and contained so that they are isolated from the human environment until the radioactivity has decayed to acceptable levels.

A major problem associated with nuclear energy is the safe disposal of high-level liquid wastes from fuel reprocessing operations. This waste contains more than 99.9 percent of the nongaseous fission products, traces of unrecovered plutonium and uranium, and the transuranic elements (so-called because, having more protons in their nuclei than uranium, they lie beyond uranium in the atomic scale; they do not exist in nature) generated in power reactors and contained originally in the spent nuclear fuel. At present, these liquid wastes are usually concentrated by evaporation and stored as an aqueous nitric acid solution in high-integrity, stainless-steel tanks. There is now general consensus within the nuclear industry that, for the long term, these liquid wastes should be converted to solid form and subsequently buried deep within the earth. Much research is in progress to develop techniques of solidification which will withstand leaching by circulating groundwater (and consequently the possible return of highly toxic radionuclides to the biosphere).

Liquid storage can be considered as an interim step between reprocessing and solidification, but the time scale and duration of liquid storage is dependent upon many factors. The irradiated fuel can be reprocessed as early as possible after discharge from the reactor if it is essential to recycle the plutonium and uranium (following a cooling period, e.g., of about 125 days for Magnox reactors). Alternatively, reprocessing of the fuel can be delayed for some years. This would enable many of the short-half-life fission products to decay and the specific activity would be lower, reducing the problems of reprocessing. The eventual solidification of the liquid waste is appreciably easier if the fission products have been cooled as long as possible. An increased length of storage, however, implies larger storage volumes and correspondingly larger capital and operating costs. The length of liquid storage periods will generally be determined by these factors and by national policies regarding the solidification and disposal of the waste.

There are two main reasons why the solidification and ultimate disposal (in deep geologic repositories) of high-level waste has not yet been demonstrated. The first is that the present and projected volumes of commercial high-level wastes for a number of decades are so small that the need does not yet exist. The second reason partly stems from the first. From a technical standpoint, the time is being used to research, develop, and evaluate alternative systems for long-term disposal of the waste.

Thirty years is generally specified for the economic life of a nuclear-power plant commissioned during the 1960s and 1970s. All plants have a useful life which is determined by many interacting economic and technical factors, and provision for decommissioning should preferably be considered at the design stage. Whether decommissioning involves immediate dismantling of the entire structure (i.e., "green field" state) or mothballing, both methods require storage, decontamination, and entombment to ensure continued protection from the residual radioactivity and other potential hazards in the retired facility. Dismantling a nuclear-power plant is likely to be a very complicated and expensive method of disposal. Entombment (in concrete) appears to be a financially more acceptable option, although this will involve the provision of adequate security over a considerable time period.

NUCLEAR FUEL CYCLE COSTS

The complexity of nuclear fuel cycle economics stems from the fact that it involves numerous expenditures made at different points in time before the fuel is actually loaded into the reactor and energy production begins, as well as other disbursements made a long time after the spent fuel has been unloaded from the reactor for ultimate disposal or for reprocessing, production of new fuel with recovered fissionable materials, and disposal of radioactive wastes.

The front-end processes of the nuclear fuel cycle include the costs incurred in the exploration, mining, and milling of uranium, conversion into UF_6, enrichment (if required), and fabrication of fuel elements. In addition, all costs for transport between processes and dispatch to the reactor site must also be included.

The back-end processes of the nuclear fuel cycle include the expenditures incurred in storage and transport of irradiated fuel, and reprocessing for extraction of plutonium and uranium, and the separation, concentration, and final disposal of radioactive wastes in the case of a closed fuel cycle. The economic effect of recycling the recovered plutonium and uranium is to add a credit to the nuclear fuel cycle costs. Interest on the expenditures incurred throughout the fuel cycle constitute the "indirect" cost of the nuclear fuel cycle. Direct plus indirect costs determine the total nuclear fuel cycle cost component of the energy produced by the fuel burned in the reactor. Since many fuel batches of different composition may be used during the life of the reactor, it is customary to calculate the levelized cost of the energy produced by the nuclear plant throughout its lifetime.

Table 2.2 contains the reference data used by the IAEA for evaluating the relative costs of fuel for alternative modes of

TABLE 2.2 Reference Fuel Cost Data (constant, 1980 U.S.$)

Fuel		Range	Reference Value
Nuclear			
Natural uranium	\$/kg U_3O_8	48–120	88
Conversion to UF_6 (LWR)	\$/kg U	4–6	5
Enrichment (LWR)	\$/SWU	120–200	160
Fabrication (LWR)	\$/kg U	150–200	175
Fabrication (HWR)	\$/kg U	80–100	85
Shipping	\$/kg U	10–20	15
Back-end cost (net)	\$/kg U	300–500	400
Discount rate (% per annum)		8–14	10
Annual load factor (%)		60–80	70
Total fuel cycle cost (LWR)	10^{-3} U.S.\$/Kwh	9.5–10.5	10
Total fuel cycle cost (HWR)	10^{-3} U.S.\$/Kwh	5–7	6
Fossil			
Fuel cost; total			
Coal (30–60–90 \$/tonne)		12–24–36	
Oil (30–40–50 \$/bbl)		44–58–72	

Source: International Atomic Energy Agency, *Guidebook on the Introduction of Nuclear Power*, IAEA, Vienna, 1982a.

electricity generation. With respect to fuel alone (i.e., ignoring construction, operating, and debt financing costs) it is apparent that uranium is an extremely small component in the cost of generating electricity relative to its fossil fuel counterparts. It was noted in Chapter 1, however, that the cost of nuclear power is considerably more volatile with respect to construction lead times and interest rates than either coal- or oil-fired electricity generation.

NOTES

1. The change in fuel composition after irradiation and fissioning within the reactor has been detailed by Rybalchenko and Colton (1981).
2. Perhaps the most thorough discussion of this issue is contained in the Flowers Report (1976). A more recent study of relevance is by Fareeduddin and Hirling (1983).

3

History of the Uranium Market

INTRODUCTION

The foundations of the modern-day uranium-mining industry were laid during the Manhattan Project and the subsequent, postwar Cold War period, which extended into the late 1950s. During these years the U.S. industry was subject to a large degree of government manipulation and regulation designed at ensuring adequate supplies for the nuclear arms race, especially in the wake of the Soviet Union's successful detonation of an atomic bomb in 1949. The United Kingdom and, a few years later, France were also developing separate nuclear capabilities and consequently required large amounts of uranium to satisfy their weapons programs. Until the mid-1960s, therefore, uranium was used almost exclusively for the construction of nuclear weapons. As stockpiles grew and world tension eased, however, the demand for uranium declined rapidly and the uranium mining industry fell into recession.

In the late-1960s, a small private market for uranium came into existence to supply the fledgling nuclear-power industry. When it became apparent that nuclear energy was a viable economic proposition, the nuclear-power industry expanded rapidly in many industrialized nations and a renewed interest in uranium production became evident. A revival of the uranium industry was assured. While direct government control over the industry was less stringent than during the weapons period, nevertheless a high degree of government involvement and regulation was (and still is) apparent at all stages of the nuclear fuel cycle.

In this chapter the uranium market's turbulent history over these two distinct periods of industry development are considered separately. The dividing year is taken as 1968, the year a commercial market for uranium was established in the United States.

THE MARKET TO 1968

The rapid wartime development of nuclear technology generated by the Manhattan Project demonstrated an urgent need for substantial supplies of uranium. From 1940 to 1945, U.S. government purchases of uranium for the development of the atom bomb came from three sources: the Shinkolobwe mine in the Belgian Congo, the Eldorado Gold Mines operation on Great Bear Lake, Canada, and the vanadium mines on the Colorado Plateau. However, these sources of supply were only capable of providing a minute proportion of the anticipated postwar requirements of the United Kingdom and the United States. Following the cessation of hostilities, therefore, a worldwide search for uranium was actively encouraged by the U.S. and U.K. governments in order to satisfy their growing demand for the raw material necessary for the expansion of their nuclear arsenals. Attention was focused on countries where uranium occurrences were known and where a favorable political climate would ensure their expeditious development. While the world's major supplier of uranium during the immediate postwar period was the Belgian Congo (the Shinkolobwe mine was reopened in 1945), exploration and development effort was centered on Canada, South Africa, the United States, and, to a lesser extent, Australia.

In 1944 the U.K. and U.S. governments formed the Combined Development Agency (CDA) as a uranium-purchasing agency charged with the responsibility of ensuring that adequate supplies of foreign uranium were obtained for the weapons programs of the two governments. Exploration for uranium and the development of both new and existing deposits were encouraged by loans, technical assistance, and a guarantee that all production would be purchased at favorable prices. In addition, many countries offered cash rewards to private individuals or companies for the discovery of new deposits.

In the mid-1950s France also became actively engaged in the search for uranium, as it sought to develop an independent nuclear weapons capability, but was able to obtain most of its requirements from domestic sources and its (then) colonies of Gabon and Madagascar.

The generous incentives for uranium discovery and production instituted by the CDA and the U.S. Atomic Energy Commission (AEC) led to a dramatic growth in the industry during the mid-1950s. In 1950 the Belgian Congo was the world's largest producer of uranium (the majority of which was exported to the United States), but this small amount of production from the Shinkolobwe deposit was soon swamped by the rapid development of production facilities in Canada, South Africa (as a by-product of gold mining), and the United States. In the second half of the 1950s, world production increased by a factor of three (see Table 3.1), as uranium rose from being a commodity of only minor commercial value to one of the most sought after minerals on earth.

In the United States, major exploration efforts by the AEC and private industry resulted in the discovery of substantial deposits in new producing areas. Over the period 1949–61, minimum base prices for ores of various types and grades were guaranteed under the AEC domestic uranium procurement program, together with initial production bonuses and quantity and development allowances. Over this period the average price paid for uranium concentrate by the AEC ranged from \$8.53/lb U_3O_8 to \$12.51/lb U_3O_8 (Table 3.2), which was an attractive enough proposition to stimulate U.S. production to reach 17,640 tons U_3O_8 in 1960 (Table 3.1), the largest figure recorded until 1978. Cost figures are difficult to obtain, but Lecraw notes that in 1972 "the producers agreed that \$6.25/lb U_3O_8 would cover all costs plus interest and amortization plus profits plus a risk premium."[1] Deflating this figure by the "all commodities" U.S. Producer Price Index gives, albeit imperfect, cost figures of \$4.61, \$4.98, and \$5.07 for 1955, 1960, and 1965, respectively. Thus the prices being offered were extremely attractive.

By the mid-1950s it had become apparent that the worldwide search for uranium deposits had been extremely successful, and that production would soon exceed requirements for military purposes. In May 1956 the AEC announced that from 1962 until the end of 1966 procurements would be at a fixed \$8/lb U_3O_8 and that ore price guarantees would end. This can be interpreted as the initial move by the AEC to dampen uranium production to avoid the buildup of an excessive stockpile. Two years later more positive steps to discourage superfluous production were taken when the AEC announced that it would not purchase concentrates produced from domestic reserves developed after 1958. Since the AEC was the only purchaser of U.S. uranium concentrate, this effectively eliminated incentives for exploration efforts, which began to decline immediately. The boom was over.

AEC domestic concentrate procurements remained relatively stable until the revised procurement program was introduced in 1962. Between 1962 and 1965, however, they declined by 35 percent. Figure 3.1 shows annual levels of AEC uranium purchases between 1956 and 1971 by both value and volume. Note the importance of Canada as a supplier over the period 1958–63 and its rapid decline thereafter. The overseas category is dominated by South Africa.

All AEC contracts for the purchase of uranium concentrate were scheduled for completion at the end of 1966, with no plans for additional purchases in subsequent years. To prevent the virtual shutdown of the entire domestic uranium mining industry, a "stretchout" program was introduced to extend the deadline for uranium concentrate purchases to the end of 1970, by which time it was anticipated that a substantial commercial market would be developing. The stretchout policy allowed for deliveries scheduled for the period 1963–66 to be deferred until 1967–68 (at the same price), with the incentive being additional AEC purchases in 1969–70 of an amount equal to that deferred. Prices for the 1969–70 purchases were fixed by a formula which allowed 85 percent of the allowable production costs (over the period 1963–68) per pound plus an extra $1.60, subject to a maximum of $6.70/lb U_3O_8. This facility was not extended to foreign producers, from whom all deliveries ceased in 1967. No uranium was purchased by the AEC after 1970, although a few domestic deliveries were allowed to be postponed until 1971.

The decline in AEC purchases of uranium had a more pronounced impact on the Canadian uranium-mining industry than its U.S. counterpart. In 1966, the AEC decided that the enrichment of foreign uranium for use in domestic reactors would threaten the viability of the domestic uranium industry, and it therefore invoked an amended section of the Atomic Energy Act of 1954 effectively prohibiting U.S. power companies from using imported uranium. Stockpiling and forward buying of foreign uranium in anticipation that the embargo would be lifted were not outlawed. Since the United States accounted for approximately 70 percent of the world's annual demand for uranium, the Canadian mining industry was severely affected. In 1959 it accounted for one-third of world production of uranium, which was valued at

TABLE 3.1 World Production of Uranium Concentrate (thousand tons U_3O_8)

	World[1]	U.S.	Canada	S. Africa	Namibia	Niger	France	Australia	Gabon[2]
1983[e]	47.84	10.60	9.40	7.93	4.90	4.42	3.77	4.18	1.30
1982	53.73	13.43	10.50	7.56	4.90	4.60	3.17	5.82	1.20
1981	57.36	19.24	10.07	7.97	5.16	5.67	3.32	3.70	1.33
1980	57.30	21.85	9.16	7.99	5.25	5.37	3.42	2.03	1.35
1979	50.02	18.73	8.86	6.24	4.98	4.71	3.07	0.92	1.46
1978	44.38	18.49	8.84	5.15	3.50	2.68	2.83	0.67	1.17
1977	36.96	14.94	7.74	4.27	3.04	2.09	2.73	0.46	1.18
1976	30.10	12.75	6.68	3.59	0.77	1.90	2.36	0.47	1.20
1975	26.44	11.60	6.13	3.10		1.82	2.23	0.00	1.21
1974	24.58	11.53	4.80	3.39		1.43	2.13	0.00	1.00
1973	25.80	13.24	4.76	3.41		1.23	1.98	0.00	0.85
1972	25.65	12.90	4.88	4.00		0.96	1.94	0.00	0.58
1971	23.91	12.27	4.11	4.19		0.47	1.94	0.00	0.60
1970	24.16	12.91	4.10	4.12		0.04	1.94	0.33	0.42
1969	23.08	11.61	3.86	3.98			1.77	0.33	0.54
1968	23.01	12.37	3.70	3.88			1.63	0.33	0.59
1967	19.10	11.25	3.74	3.36			1.59	0.33	0.63
1966	19.52	10.59	3.93	3.29			1.54	0.33	0.62
1965	20.59	10.44	4.44	2.94			1.42	0.37	0.72
1964	26.20	11.85	7.29	4.45			1.33	0.37	0.59
1963	31.03	14.22	8.35	4.53			1.99	1.20	0.58

1962	34.50	17.01	8.43	5.02			1.98	1.30	0.51
1961	36.30	17.35	9.64	5.47	Zaire[3]		1.62	1.40	0.43
1960	41.13	17.64	12.75	6.41	1.20		1.38	1.30	
1959	43.35	16.24	15.89	6.45	2.30		0.95	1.10	
1958	36.25	12.44	13.40	6.25	2.30		0.66	0.70	
1957	23.27	8.48	6.64	5.70	1.30		0.47	0.40	
1956	14.50	5.96	2.28	4.40	1.30		0.26	0.30	
1955	8.55	2.78	1.26	2.80	1.25		0.23	0.23	
1954	6.09	1.70	1.26	1.68	1.25		0.10	0.10	
1953	4.55	1.20	1.00	1.00	1.25		0.10		
1952	3.20	1.00	0.60	0.10	1.40		0.10		
1951	3.35	0.90	0.25		2.10		0.10		
1950	3.55	0.55	0.25		2.65		0.10		
1949	2.50	0.50	0.10		1.80		<0.10		
1948	2.10	0.30	0.10		1.70				

[1] Prior to 1961, estimates were added to the World total to allow for production from some minor uranium-producing nations. Between 1961 and 1976 the World figure is simply the sum of the individual nations appearing in the table, together with minor levels of output from Argentina, Portugal, Spain, and Sweden. After 1976, the production of all noncentrally planned economies is included in the World figure where possible.

[2] Prior to 1961, the Gabon figure was included with France.

[3] Production in Zaire (then the Belgian Congo) was discontinued in 1961.

[e] Estimated values for World, Namibia, Niger, and Gabon.

Sources: U.S. data: *Statistical Data of the Uranium Industry,* U.S. Department of Energy, Grand Junction GJO — 100(82), January 1, 1983. Other data: up to 1976, *Minerals Yearbook,* U.S. Department of Mines, various issues; after 1976: *Australian Mineral Industry Annual Review,* Australian Government Publishing Service, Canberra, various issues.

TABLE 3.2 Uranium Prices[1] (\$/lb U_3O_8) (unweighted yearly averages)

	Current Prices	Constant (1967) Prices		Current Prices	Constant (1967) Prices
1949	8.53	11.34	1967	8.00	8.00
1950	8.92	11.44	1968	7.98	7.79
1951	10.01	11.63	1969	6.20	5.85
1952	11.19	13.31	1970	6.24	5.67
1953	12.30	14.52	1971	6.06	5.32
1954	12.25	14.39	1972	5.95	5.05
1955	12.51	14.40	1973	6.41	5.09
1956	11.63	12.81	1974	11.03	7.17
1957	10.53	11.29	1975	23.68	13.81
1958	9.57	10.22	1976	39.70	21.77
1959	9.40	9.87	1977	42.20	21.63
1960	8.99	9.43	1978	43.23	20.64
1961	8.54	9.01	1979	42.57	18.00
1962	8.00	8.44	1980	31.79	11.57
1963	8.00	8.45	1981	24.19	7.95
1964	8.00	8.40	1982	19.90	6.37
1965	8.00	8.30	1983	22.98	7.28
1966	8.00	8.12	1984	17.27	5.35

Note: For the series in constant prices, the current prices were deflated by the U.S. Producer Price Index (industrial commodities) with 1967 as base year.

[1] Prior to 1962 the reported price is the average cost per pound of U.S. purchases of concentrate (both domestic and imported). From 1962 to 1968 it is the fixed price paid for AEC procurements, and from 1969 onwards it is NUEXCO's exchange value.

Sources: NUEXCO, Monthly Report on the Nuclear Fuel Market, January 1984. *Metal Statistics,* various issues.

C\$310 million (about 6 percent of total Canadian exports). Despite Canadian government stretchout and stockpiling programs, many Canadian mines were closed and production and exploration were drastically curtailed. By 1968 production had fallen to 3,700 tons U_3O_8 compared with 15,890 tons U_3O_8 just nine years earlier. Production in Australia and South Africa was also severely cut back but with less serious consequences than experienced by Canada. Most of the Australian deposits being mined at this time were near depletion anyway. South African uranium is recovered as a by-product of gold production, and thus the major industry was unaffected. The uranium market in the late 1960s was characterized by large stockpiles, worldwide overproduction, and conse-

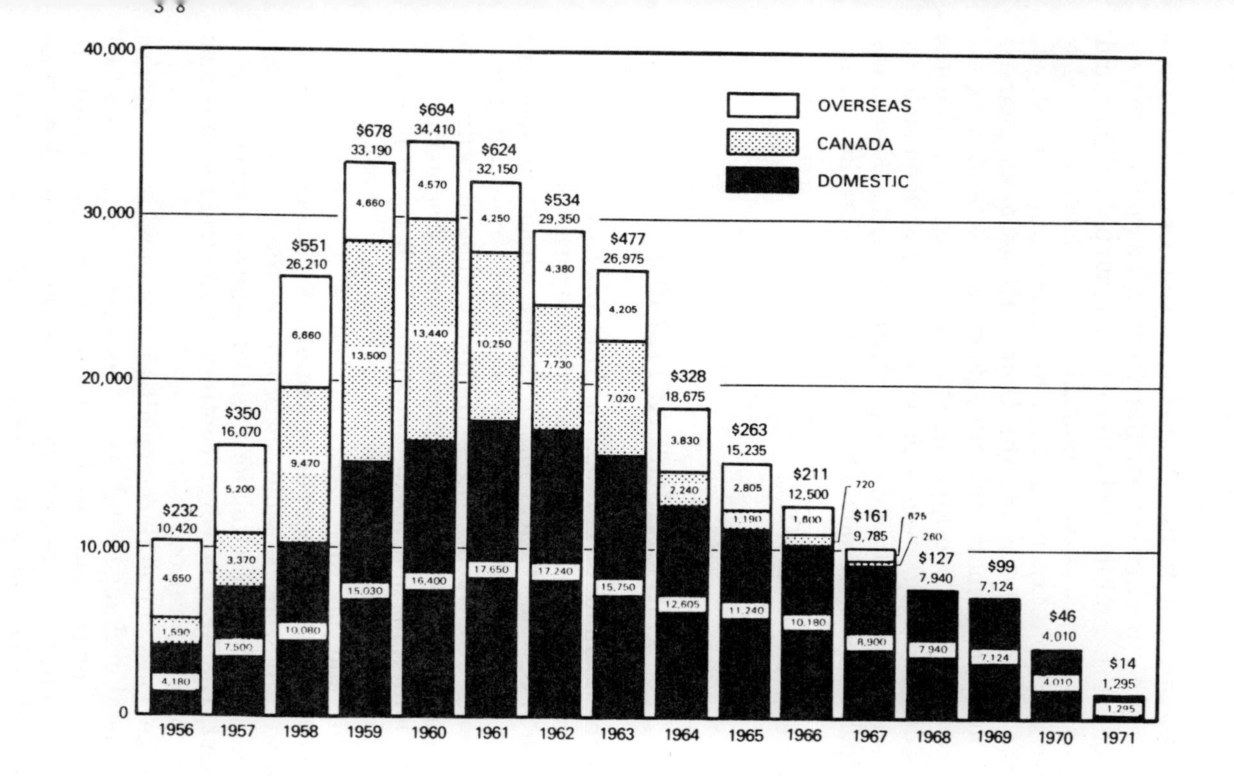

FIGURE 3.1 AEC Uranium Purchases (tons U_3O_8 and millions of dollars).

quently a reluctance by utilities to enter into long-term supply contracts.

For the period up to 1968, therefore, the prices given in Table 3.2 do not represent a market equilibrium price as would have been determined by the unhindered interaction of supply and demand. During the 1950s prices were maintained at a level designed to encourage the rapid expansion of the uranium-mining industry to enable the United States and the United Kingdom to establish adequate nuclear arsenals and strategic stockpiles. Following the end of the Cold War, stockpiling, stretchout programs, and the growth of demand from power utilities, combined with an embargo on uranium imports, sheltered U.S. producers from the uneconomic prices which competition would have determined. Certainly during this period non-U.S. producers were selling at prices considerably below those given in Table 3.2.

THE MARKET AFTER 1968

Commencing on January 1, 1969, the AEC was permitted to begin enriching privately owned uranium under the terms of the Private Ownership of Special Nuclear Materials Act, which was approved in August 1964. Enrichment services were provided under "requirements" contracts. These contracts allowed utilities to specify their separative work requirements on a short-term basis, subject to minimum and maximum limits. Both the transaction tails assay and operating tails assay were set at 0.20 percent.

At the end of 1970, AEC purchases of uranium were discontinued with its stockpile standing at 50,000 tons U_3O_8. Utilities were now responsible for procuring their own uranium requirements and contracting for their subsequent conversion and enrichment. The AEC, however, continued to control the provision of all enrichment services.

Unlike most mineral commodities, uranium is not traded through metal exchanges and, as a consequence, there is no spot price for uranium transactions. In addition, information relating to prices for uranium concentrate deliveries under individual long-term private market contracts is not generally made public, nor are the terms of the contracts usually revealed. Since 1968, however, NUEXCO, the world's principal private uranium brokerage company, has published a monthly "exchange value" price for uranium concentrate. This value represents NUEXCO's "judgement of the price at which transactions for significant quantities of natural uranium concentrate could be concluded as of the last day of the month." They do not consider old or renegotiated contracts. Exchange values, therefore, are current prices for current or short-term delivery. While there have been minor changes over time in NUEXCO's method of calculating exchange values, and despite the fact that NUEXCO emphasizes that it is not a spot price in the usual sense of the word, it is nevertheless generally regarded as an indicator of uranium market price levels. In a market where long-term contracts are now commonplace, the spot price serves to reflect short-run variations in the conditions of supply and demand. Deliveries on spot or near-term (i.e., within one year) contracts typically account for about 10–20 percent of total deliveries. The prices given in Table 3.2 for 1968 onward are NUEXCO exchange values.

Although the nuclear-power industry entered a period of rapid expansion after 1965, this growth had no immediate impact on the demand for uranium. At the end of 1968 there was just 11.9 GWe of operational nuclear capacity worldwide (including the centrally planned economies), with 80.1 GWe of capacity "under construction" and 34.6 GWe of capacity planned. Electricity consumption in the developed countries approximately doubled in the 1960s, and power utilities were investing heavily in additional generating capacity in the belief that this rate of growth would continue unchecked for the foreseeable future. A central role was foreseen for nuclear power over the remaining years of the twentieth century, with the level of operational capacity projected to reach 300 GWe in WOCA countries by 1980. The raw materials to fuel this projected growth in power consumption (coal, oil, and uranium) were all being supplied at (what appears in retrospect to have been) bargain prices, and a prosperous 1970s seemed assured.

There is a considerable time lag, however, between the date of ordering and the date of commercial operation of a nuclear-powered plant (ranging from about six to ten years in the early 1970s) due not only to the large-scale nature of the construction project itself but also to the time taken to comply with licensing and regulatory requirements before commercial operations can commence. Thus the expected growth in demand for uranium from the power utilities could not have been expected to occur much before the mid-1970s. Over the period 1968–72, therefore, prices continued to fall (in real terms the fall was considerable) as a result of excessive stocks, overproduction, and delays in the licensing and construction of nuclear power plants.

Outside the United States, the problem was far more acute, particularly in Canada. Lecraw quotes an instance where a Canadian producer lost a bid at \$3.95/lb U_3O_8 in 1971![2] The depressed state of the industry led to a meeting in Paris in 1972 of representatives from the British-based mining company Rio Tinto Zinc and government representatives from Australia, Canada, France, and South Africa to form a Uranium Marketing Research Organization (often referred to as "The Club"). The precise nature of the club's activities, however, was exposed in 1976, when Friends of the Earth in Australia "acquired" and made public copies of secret memos and letters from the files of Mary Kathleen Uranium. This stolen material documented the existence of the Club's role as a cartel which set price schedules and regulations, quotas, and terms and conditions of sale for all transactions. Members who broke the cartel rules were penalized in cash, tonnage, and quotas. Although club member countries and the United States were specifically exempted from conditions fixed by the cartel agreement, it appears that U.S. uranium producers demanded prices that were at least as high as those set by the cartel.

Until 1973 uranium-supply contracts were typically short-term in nature, reflecting the depressed state of the uranium-mining industry. Prices had stagnated for almost a decade and prospects for any recovery in the U.S. market were overshadowed by the AEC's 50,000 tons U_3O_8 stockpile which, it was generally thought, would be sold on the private market as soon as prices recovered. In addition, a major manufacturer of nuclear reactors, the Westinghouse Electric Corporation, had guaranteed to supply a number of power utilities with their uranium requirements at the current (depressed) price over their early years of operation as an inducement to purchase Westinghouse reactors. Thus there was little incentive for utilities to enter into long-term purchase contracts with uranium producers. The uranium market, however, was about to experience a very spectacular and extremely controversial, but fairly short-lived, revival.

The Initial Price Shock

In order to ensure full utilization of its enrichment plants, in May 1973 the AEC changed its enrichment contracting procedures. Henceforth, under the conditions of the Long Term Fixed-Commitment Contract (LTFC), utilities (both domestic and foreign) would be obliged to commit themselves to long-term contracts for the supply of their enrichment requirements (i.e., separative work), thereby also inducing them to purchase uranium on a long-term planning basis. A December 31, 1973, deadline was set for execution of contracts for initial delivery of enriched uranium before July 1, 1978, while a deadline of June 30, 1974, was set for contracts for initial delivery of enriched uranium between July 1, 1978, and June 30, 1982. In addition, utilities were obliged to commit themselves for an amount of separative work needed to cover the first ten-year period of enriched uranium deliveries. In effect, this meant that utilities had to plan and to commit themselves for a period of up to 18 years. Being conservative by nature, utilities tended to err on the safe side, ordering in excess of their theoretical requirements. Due to the lead time of up to eight years imposed by the second deadline, enrichment contracts were concluded even for nuclear-power stations which were only in the preliminary planning stage. Just one year later, all AEC enrichment capacity was under contract and the order books were closed.

Any reservations that utilities might have had over the wisdom of entering enrichment supply contracts over such a long time horizon were soon dismissed when a few months later the OPEC oil embargo and spiraling oil prices cast a shadow of an impending world energy shortage.

The world energy arena during the mid-1970s was dominated by the aggressive oil pricing and marketing policies pursued by the nations belonging to OPEC. For the previous 20 years, the world's giant oil companies had purchased Middle East oil at a price which had varied little in nominal terms and had thus sunk substantially in real terms. As a consequence, the economies of most industrialized nations had become heavily dependent upon imported supplies of this relatively cheap source of energy. In 1972, oil accounted for 55 percent of the overall energy consumption of OECD nations, with coal, natural gas, hydro-geothermal, and nuclear power having shares of 19 percent, 22 percent, 3 percent, and 1 percent, respectively. In the case of Japan, 77 percent of total energy consumption was in the form of oil.

Faced with increasing oil demand, falling reserves, and generally backward domestic economies, the more aggressive members of OPEC felt that oil revenues should be raised in an attempt to achieve industrial and social advancement. Following a period of unsuccessful negotiations regarding the extent of a proposed increase in the price of oil between OPEC member states and representatives of the major oil companies, on October 13, 1973, OPEC unilaterally raised the posted price of Saudi Arabian light oil (the reference price for all other oil) by 77 percent, from $2.90 to $5.12 per barrel.

Exactly a week earlier, war had broken out between Israel and two of its Arab neighbors, Egypt and Syria. After early setbacks, the successes of the Israeli forces were numerous and swift. In an attempt to dissuade many Western nations from openly supporting Israel, OPEC placed an embargo on oil exports to the Netherlands and the United States, while other "pro-Israeli" countries had deliveries cut by an initial 10 percent, followed by 5 percent a month until what OPEC felt were the Arab goals had been achieved. Although hostilities only lasted 18 days, the embargo was not removed until mid-March, 1974.

Over a period of less than three weeks, therefore, the international oil market had undergone a radical and permanent change from the dominance of a consumer-led cartel (oil companies) to that of a producer-led cartel. Rumors of massive oil price hikes in OPEC's posted price were commonplace in late 1973, and eventually OPEC decided on a new posted price of $11.65 per barrel, effective January 1, 1974. Thus the price had risen fourfold in just 10 weeks.

The awesome power of OPEC in demanding and obtaining these higher prices raised fears for the economic viability of Western industrialized nations. Speculative behavior led to rapid increases in the prices of numerous strategic commodities,[3] stimulated by near-panic inventory accumulation, especially following thinly veiled threats of U.S. military intervention to smash OPEC if the U.S. economy came under danger of collapse from "exorbitant" increases in the price of oil.

Uranium prices responded in parallel to other raw material prices. As a result of the combined impact of the LTFC and the oil embargo and price hikes, U.S. and foreign utilities rushed into long-term supply contracts in order to ensure the security of their future uranium requirements. Just a year earlier, the AEC had decided against a proposal to sell its uranium stockpile on the open market in favor of operating its enrichment plants under a "split-tails" policy.[4] This decision, in effect, allowed for the disposal of the stockpile without affecting sales of uranium producers. The immediate impact of this sudden substantial increase in demand for uranium, combined with a relatively static level of supply from a depressed mining industry, was to generate a substantial rise in the spot price. Since the AEC stockpile was being disposed of through the enrichment process, and with the embargo on the domestic use of foreign uranium, the potential for a substantial increase in the supply of uranium to the U.S. market was very limited. In nominal terms, the spot price of uranium rose by a factor of four (from \$6.50/lb U_3O_8 to \$26/lb U_3O_8) between October 1973 and August 1975.

Since virtually all industrialized nations of the world were at that time net importers of oil, an increased emphasis on nuclear power was envisaged by many nations as a way of reducing their dependence on the vagaries of the oil pricing policies of potentially unfriendly nations. Even the removal of the embargo, an oil price freeze, and the lifting of OPEC production levels in mid-March 1974 did little to change the shift in favor toward nuclear power.

However, the very long lead times which are a characteristic of investment and plant construction procedures in the electricity supply industry restricted the speed with which a transition from oil-fired to alternative forms of electricity-generating plant could be made. By the end of 1982, nine years after the OPEC embargo and three years after a further substantial increase in oil prices, oil still accounted for 45 percent of overall energy use in OECD nations, although nuclear power's share had risen to 5 percent. In the short term, following the initial price increases, the only possibility for industrialized nations to reduce their dependence on high-cost imports of oil was to shift as much of the load as possible to their coal-fired (or other alternative) plants and to engage in extensive programs aimed at energy conservation.

If the price of a commodity rises then, ceteris paribus, the price of competing commodities will also rise in response to the short-term increase in demand brought about by consumers shifting their buying patterns to incorporate a greater quantity of the relatively cheaper substitute. Suppliers of the substitute commodity stand to make windfall profits (called quasi-rent in economic theory),[5] since in the short run their revenue exceeds that which was required to induce them to produce the substitute. In a competitive market, however, these gains will disappear in the long run as additional supplies, attracted by the high prices, will eventually be produced.[6] In the case of electricity generation, coal and uranium can be substituted for oil in the short run if excess generating capacity allows for a shift of load away from the oil-fired plants in favor of coal-fired and nuclear-powered plants. Over a somewhat longer period, additional coal or nuclear capacity can be installed and existing oil-fired plants retired. It was not surprising, therefore, that given the magnitude of the increase in oil prices in 1973 substantial increases were also recorded in the prices of competing commodities.

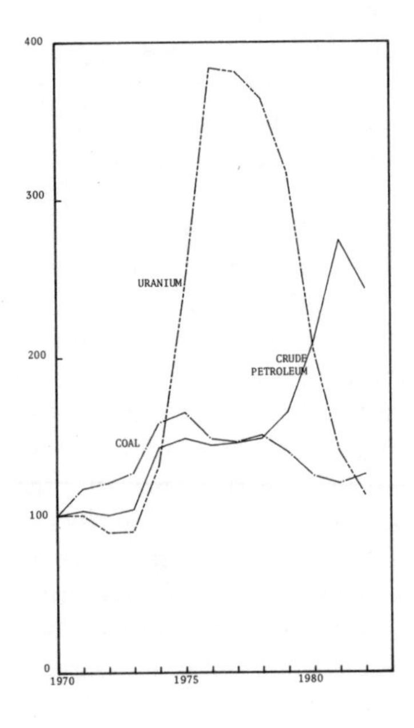

FIGURE 3.2 U.S. Price Indices for Coal, Crude Petroleum, and Uranium (unweighted yearly averages, 1970 = 100).
Source: NUEXCO, Monthly Report on the Nuclear Fuel Market, October 1983. *Monthly Labor Review,* U.S. Bureau of Labor Statistics.

Movements since 1969 in U.S. spot prices (in real terms) for coal, crude petroleum, and uranium are illustrated in Figure 3.2. While these indexes give a picture of general price movements in competing fuels, they are not strictly comparable, due to different definitions of the spot price and different relative levels of importance of the spot market in the industries considered. It should also be noted that the ultimate effect on the price of electricity of similar increases in the price of these three fuels is far greater in the case of the fossil fuels than it is for uranium.[7] It is also relevant to remember that Western European nations without sizable domestic supplies of oil would have faced considerably higher spot prices for oil, buying either on the Rotterdam spot market or from the Middle East, than experienced by the United States over this period, as the latter's domestic supplies tempered the price increases.

By the first half of 1975, it is apparent from Figure 3.2 that coal and crude petroleum prices were beginning to stabilize (in real terms). While uranium prices also experienced a short period of stability around the middle of 1975, a further surge in prices was about to occur.

In an attempt to relieve the domestic uranium market of excessive demand-side pressure created as a result of the LTFC and the OPEC crisis, in October 1974 the AEC announced that the restrictions on the enrichment of foreign uranium for domestic use would gradually be lifted over the period 1977–83. The amount of foreign uranium feed which any domestic customer delivered to the enrichment plants could not exceed 10 percent of his total deliveries in 1977, 15 percent in 1978, 20 percent in 1979, 30 percent in 1980, 40 percent in 1981, 60 percent in 1982, and 80 percent in 1983. Thereafter no restriction applied.

By mid-1975 the price of uranium had begun to stabilize. The aftermath of the oil price rises of 1973 and 1974 had been a drop in the level of industrial activity in most industrialized nations and an increased emphasis on measures for fuel conservation. As a result, the rate of growth of the demand for electricity slackened and, as new power plants came on-stream, excess capacity started to build up. Far from experiencing a boom at the expense of the oil industry, nuclear power was encountering problems of its own. Delays in the construction and licensing of new reactors combined with exaggerated ordering of enrichment services by utilities in 1973 forced ERDA (Energy Research and Development Administration), the AEC's successor from January 1975 in the provision of enrichment services, to release some of the demand-side pressure on the uranium market by allowing a two-month "open season" during which contracts for enrichment services could be adjusted or terminated.

To this point we have only considered demand-side factors in explaining the 1973–75 surge in the price of uranium. While the U.S. embargo on the enrichment of foreign uranium effectively restricted the supply of uranium to the U.S. market, any influence that the uranium cartel may have exerted on prices has largely been ignored. In September 1975, however, the uranium market was hit by an event that was to bring the cartel widespread publicity and its members a protracted and expensive legal battle.

The Uranium Cartel and the Westinghouse Case[8]

There is no question that an international uranium producers cartel, composed of representatives from Australia, Canada, France, and South Africa, met a number of times between February

1972 and May 1974 to fix prices for uranium concentrate. Since the production of uranium in these four countries (together with the then-French colonies of Gabon and Niger) is subject to a significant degree of domestic government control, one can assume that the formation of a cartel received at least tacit support from the governments of these countries. The basic aim of the cartel was to establish a "floor price" for uranium exports and end a period of price warfare brought about by excessive supplies of uranium flooding the very small, non-U.S., market. At that time, the U.S. embargo on imported uranium effectively prohibited cartel members from selling uranium to U.S. utilities. The cartel ceased operations early in 1975 (by which time the price of uranium had reached \$20/lb U_3O_8), since its aim of bolstering the world price of uranium had become redundant, and most commentators have concluded that the cartel had little effect on prices.[9]

This view was not shared by the Westinghouse Electric Corporation which, following the disclosure of the existence of the cartel in the documents stolen from Mary Kathleen Uranium, immediately sought damages from 17 U.S. and 12 non-U.S. uranium companies for operating a worldwide cartel which, it claimed, conspired to bring about price rises and adopted practices that damaged Westinghouse.

Before the post-1973 price rise, Westinghouse had agreed to supply approximately 40,000 tons U_3O_8 (with deliveries ranging from immediate to 15 years) to electric utilities at the current (i.e., \$8–\$10/lb U_3O_8) prices as an added incentive for purchasing Westinghouse nuclear-power reactors. In what must rate highly as one of the major commercial blunders in U.S. history, apparently in the mistaken belief that uranium prices would remain depressed during the 1970s, they delayed making uranium-purchasing commitments sufficient to honor their contractual obligations. The post-1973 rise in uranium prices caught Westinghouse with only about 7,500 tons U_3O_8 to cover itself. Thus every rise of \$1 in the price of uranium meant an increased outlay of \$65 million if Westinghouse were to honor its obligations. By early September 1975 the price of uranium had reached \$26/lb U_3O_8, which would have involved Westinghouse in an outlay of about \$1.2 billion to obtain the necessary uranium supplies on the open market. Faced with the prospect of bankruptcy, Westinghouse declared its intention to renege on its contractual obligations, citing "commercial impracticability." It was in default by about 32,500 tons U_3O_8, approximately two-and-a-half times the total U.S. production of uranium in 1975. A number of utilities in both the United States and Sweden promptly filed suits in an attempt to force Westinghouse to comply with its contracts.

The massive legal repercussions that followed the cartel's exposure involved almost every U.S. uranium producer, 27 utilities, and two of the largest U.S. corporations, Gulf Oil and Westinghouse. Gulf Oil and its subsidiary, Gulf Minerals Canada, were at the center of the international legal disputes that the cartel case set off. After a protracted legal battle, Westinghouse and Gulf reached an out-of-court settlement in 1980, and all other parties to the dispute eventually followed suit. Basically, the producers agreed to pay cash settlements to Westinghouse and to supply it with certain quantities of uranium at reduced prices, thus enabling the company to meet, at least partly, its previously unsecured commitments.

The Second Price Shock

Following the Westinghouse decision to renege on its contracts, the utilities affected by this action were forced into the short-term market to satisfy their requirements to meet enrichment contract schedules. The spot and short-term market, however, was relatively small at that time, reflecting the long-term nature of the uranium procurement contracts entered into by utilities following the introduction of the LTFC. Consequently the pressure of a substantial increase in short-term demand initiated another price surge. Following a relatively short period of stable prices in mid-1975, the exchange value rose from \$26/lb U_3O_8 in September 1975 (Westinghouse "dropped the bomb" on September 9, 1975) to \$40/lb U_3O_8 in April 1976. Overall, therefore, the exchange value had risen more than sixfold in just 2.5 years. By the time the major case between the utilities and Westinghouse came to court in November 1977, it was claimed[10] that Westinghouse would require \$3.1 billion to acquire the uranium that would be outstanding if the contracts were not canceled.

It should be remembered that the U.S. embargo on the enrichment of foreign uranium for domestic consumption was still in force, and thus the short-term supply of uranium to the U.S. market was relatively fixed. In addition, domestic producers reacted very slowly to the post-OPEC price increases, and it was not until 1977 that U.S. production passed its 1973 level.[11] Thereafter, production certainly boomed as prices hovered around \$40/lb U_3O_8. A worldwide expansion in uranium exploration and mining activity was generated as the result of the spot prices prevailing during the 1970s, and Table 3.1 reflects the success of many producers, especially in the United States, Canada, Namibia, Niger, and, at a later date and to a lesser extent, Australia.

Before the rapid increase occurred in the exchange value, long-term uranium supply contracts typically specified fixed delivery prices. Following the price rise, however, many uranium producers attempted to renegotiate (frequently successfully) existing long-term contracts that had been written at prices far below those prevailing after 1975.[12] As contracts fell due for renewal, or as new contracts were negotiated, fixed-price contracts for future deliveries declined in importance and market price or indexed delivery prices became more popular with suppliers. For U.S. deliveries in 1983, 29 percent were made under contract-price type of procurement and 66 percent involved delivery at the prevailing "market" price. Of the market-price deliveries, however, 48 percent of all such contracts contained a floor price which effectively gave uranium suppliers a guaranteed minimum revenue for future deliveries. There has also been a trend for utilities to acquire uranium properties, thus giving them direct access to guaranteed supplies of uranium at prices which are not subject to the vagaries of the spot market.

THE SLUMP

While the exchange value continued to rise (in nominal terms) during the period 1976–78, the extent of the rise was small and in real terms a decline was evident. In the late-1970s, however, a substantial reduction occurred in the rate of growth in the demand for electricity as industrial production worldwide began to stagnate. New orders for nuclear plants dropped sharply, and as the recession continued into the 1980s many existing orders were canceled and construction on many others delayed or deferred. As a consequence, projections of uranium requirements for the remaining years of this century have been subject to substantial downward revision since the late 1970s. At the same time, however, new uranium-mining ventures (especially in Australia and Canada), which had been developed in response to the high prices prevailing in the mid-1970s, are approaching or have reached the production stage. As a result, uranium requirements are expected to remain below production until at least the late-1980s.

The nuclear-power industry has very long lead times, due not only to the large-scale nature of the construction projects themselves, but also to the regulatory and licensing controls imposed by various governments. Because of the decline in the rate of growth of demand for electricity after 1974, few new construction projects for nuclear power (or fossil-fired) plants have been started since 1979. Nevertheless, operational nuclear capacity in Western nations is projected to double between 1982 and 1990 as plants ordered during the first half of the 1970s and delayed for various reasons come on-stream. Thereafter, however, the momentum will disappear as few countries have follow-on construction programs for the 1990s. No new domestic orders have been placed in the Federal Republic of Germany since 1975 or in the United States since 1979, while expansion plans in Italy, Spain, Sweden, and the United Kingdom have been severely restricted by drawn-out environmental and safety arguments combined with reduced pressure for additional electricity supply during the industrial recession of the late 1970s/early 1980s.

Thus reactor cancellations and a lack of new orders, brought about by an anticipated surplus of electricity-generating capacity (continuing into the next decade in a number of countries), combined with excessive levels of uranium inventories in most consumer nations and public skepticism with respect to reactor safety following a well-publicized "accident" at Three Mile Island on March 28, 1979, forced the spot price of uranium down by 62.5 percent over the five-year period January 1980 to January 1985 (i.e., from \$40/lb U_3O_8 to \$15/lb U_3O_8). In real terms, January, 1985 saw the exchange value fall to its lowest level since NUEXCO began its publication in 1968. Short-term prospects for a price recovery appear to be extremely poor.

NOTES

1. Lecraw (1977).
2. Lecraw (1977).
3. A detailed consideration of the factors that contributed toward these price rises can be found in U.S. Department of Commerce (1976).
4. Details of the "split-tails" policy are given in Chapter 9.
5. A description of economic rent and quasi-rent can be found in most microeconomic textbooks; see, for example, Quirk (1976).
6. At the time of these events the world's major uranium producers (excluding U.S. producers) had formed a cartel with the expressed intention of ensuring

"adequate" prices for its members. Radetzki (1981) maintains that uranium producers operating within the cartel, as well as U.S. producers, restricted supply following the uranium price rises of the mid-1970s, thus ensuring a continuation of their windfall profits.

7. Currently, the fuel (i.e., U_3O_8) cost for nuclear-powered plants in the United States averages about 20 percent of the total cost of electricity generation. Comparable figures for coal- and oil-fired plants are 50 percent and 75 percent, respectively. Detailed comparisons can be found in Eden et al. (1981), while projected cost figures for 1995 are contained in U.S. Department of Energy (1982).
8. A comprehensive exposition of this dispute is given by Taylor and Yokell (1979). An economic analysis of the events and market conditions that encouraged the establishment of the cartel, and an interpretation of its impact on the uranium market, are given by Radetzki (1981). Joskow (1977) discusses the dispute from a legal standpoint.
9. Joskow (1977), Lecraw (1977), Taylor and Yokell (1979). Radetzki (1981) does not share this view. He maintains that U.S. uranium producers deliberately failed to increase production as prices rose and hence gave the impression of a supply shortfall. As a consequence, prices were forced to levels higher than they might otherwise have reached. Radetzki's association with the Westinghouse Corporation (refer to his "Acknowledgements") should be borne in mind when reading his analysis.
10. *San Francisco Chronicle*, November 30, 1977, p. 28, quoted in Taylor and Yokell (1979).
11. Refer to Radetzki's hypothesis in note 9.
12. Radetzki (1981), p. 27, gives a number of examples of prices being successfully renegotiated.

4

World Resources, Reserves, and Exploration

DEFINITIONS

Reports published periodically by the NEA (OECD) and the IAEA[1] provide estimates of WOCA uranium resources classified on the basis of geological knowledge and the estimated cost of exploitation. Geological knowledge is reflected in the levels of confidence in the occurrence and quantities of resources, for which purpose four resource categories are defined: Reasonably Assured Resources, Estimated Additional Resources–Category I,[2] Estimated Additional Resources–Category II, and Speculative Resources. These are further separated into three levels of exploitability based on the cost of their recovery: less than \$80/kg U (\$30/lb U_3O_8), \$80 to \$130/kg U (\$30–\$50/lb U_3O_8), and \$130 to \$260/kg U (\$50–\$100/lb U_3O_8). The costs are expressed in terms of U.S. dollars as of January 1, 1983. A diagramatic representation of this classification system is given in Figure 4.1.

OECD/IAEA definitions of these resource categories are as follows:[3]

> **Reasonably Assured Resources** (RAR) refers to uranium that occurs in known mineral deposits of such size, grade, and configuration that it could be recovered within the given production cost ranges, with currently proven mining and processing technology. Estimates of tonnage and grade are

based on specific sample data and measurements of the deposits and on knowledge of deposit characteristics. Reasonably Assured Resources have a high assurance of existence and in the cost category below $80/kg U are considered as *reserves*.

Estimated Additional Resources–Category I (EAR–I) refers to uranium in addition to RAR that is expected to occur, mostly on the basis of direct geological evidence, in extensions of well-explored deposits, and in deposits in which geological continuity has been established but where specific data and measurements of the deposits and knowledge of the deposits' characteristics are considered to be inadequate to classify the resource as RAR. Such deposits can be delineated and the uranium subsequently recovered, all within the given cost ranges. Estimates of tonnage and grade are based on such sampling as is available and on knowledge of the deposit characteristics as determined in the best-known parts of the deposit or in similar deposits. Less reliance can be placed on the estimates in this category than on those for RAR.

Estimated Additional Resources–Category II (EAR–II) refers to uranium in addition to EAR–I that is expected to occur in deposits believed to exist in well-defined geological trends or areas of mineralization with known deposits. Such deposits can be discovered, delineated and the uranium subsequently recovered, all within the given cost ranges. Estimates of tonnage and grade are based primarily on knowledge of deposit characteristics in known deposits within the respective trends or areas and on such sampling, geological, geophysical, or geochemical evidence as may be available. Less reliance can be placed on the estimates in this category than on those for EAR–I.

Speculative Resources (SR) refers to uranium, in addition to Estimated Additional Resources–Category II, that is thought to exist mostly on the basis of indirect evidence and geological extrapolations, in deposits discoverable with existing exploration techniques. The location of deposits envisaged in this category could generally be specified only as being somewhere within a given region or geological trend. As the term implies, the existence and size of such resources are highly speculative.

Estimates of Speculative Resources are, by definition, highly subjective and at best only a guide to the future. In addition, few countries will seek to evaluate resources falling in the highest cost category since their recovery is extremely unlikely in the foreseeable future. Thus the shaded area of Figure 4.1 indicates the

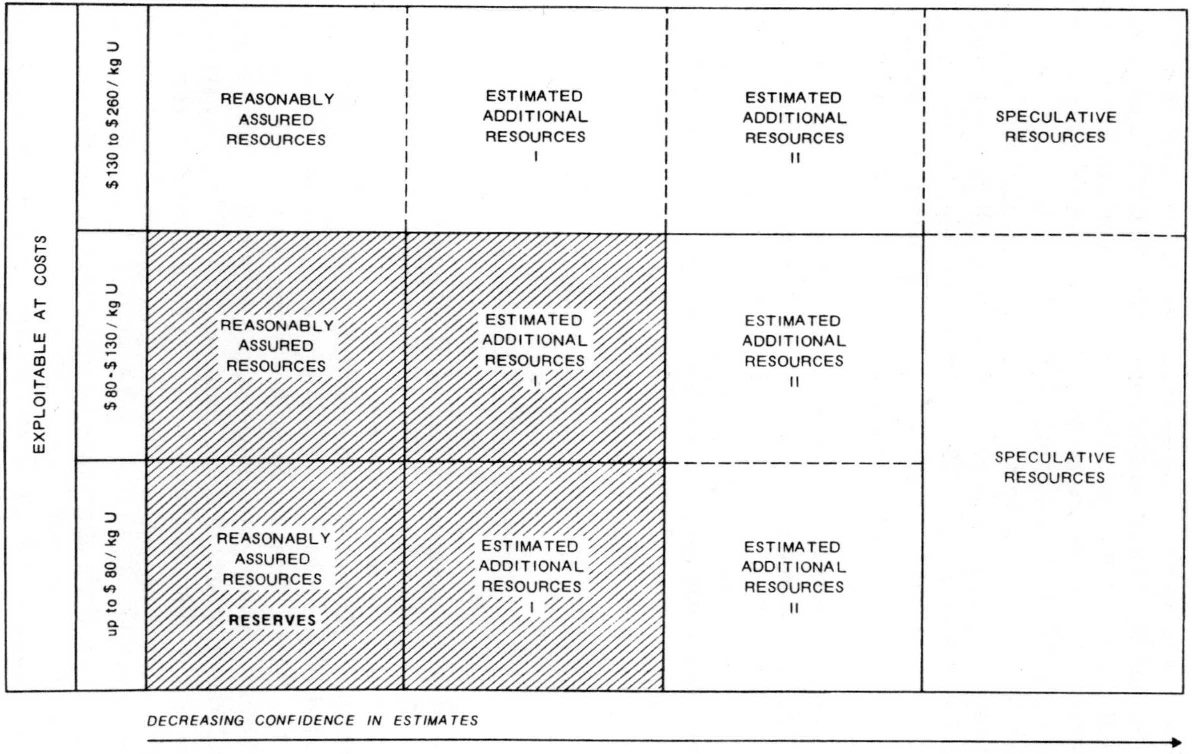

EXPLOITABLE AT COSTS				
$130 to $260 / kg U	REASONABLY ASSURED RESOURCES	ESTIMATED ADDITIONAL RESOURCES I	ESTIMATED ADDITIONAL RESOURCES II	SPECULATIVE RESOURCES
$80 - $130 / kg U	REASONABLY ASSURED RESOURCES	ESTIMATED ADDITIONAL RESOURCES I	ESTIMATED ADDITIONAL RESOURCES II	SPECULATIVE RESOURCES
up to $80 / kg U	REASONABLY ASSURED RESOURCES RESERVES	ESTIMATED ADDITIONAL RESOURCES I	ESTIMATED ADDITIONAL RESOURCES II	

DECREASING CONFIDENCE IN ESTIMATES →

FIGURE 4.1 OECD/IAEA Classification Scheme for Uranium Resources.

principal resource categories (i.e., "known resources"), while the dashed lines between RAR, EAR–I, EAR–II, and SR in the highest cost category indicate that the distinctions of level of confidence are not always applicable. Because resources in the EAR–II and SR categories are essentially undiscovered, the information on them is such that it is not always possible to divide them into different cost categories and this is indicated by the horizontal dashed lines between the different cost categories.

While minor variations arise between countries over the definition of costs, most cost figures contain all forward costs of production, including the direct costs of mining, processing, and associated environmental and waste management; costs related to maintaining existing production units and providing new units, together with an acceptable rate of return on invested capital; as well as such indirect costs as office overheads, taxes, and royalties. Property acquisition costs, past exploration and development costs, and the cost of debt servicing are generally not included. These cost categories, therefore, should not be confused with market prices, although frequently they are, since in the long run prices would be considerably higher reflecting the total long-term marginal costs of production.

RESOURCE ESTIMATES

Estimated world uranium resources (RAR and EAR–I) at January 1, 1983, are summarized in Table 4.1 for the less-than-\$30/lb U_3O_8 and \$30–\$50/lb U_3O_8 cost categories.

About 85 percent of "reserves" (i.e., <\$30/lb U_3O_8 RAR) and 94 percent of low cost EAR–I are located in just seven countries: Australia, Brazil, Canada, Namibia, Niger, South Africa, and the United States. This figure, however, probably reflects the extent of the exploration efforts in these countries, rather than an asymmetric geological distribution of the world's uranium resources. Resource estimates for the higher (i.e., \$30–\$50/lb U_3O_8) cost category must be considered very tentative. Even where known resources exist, little effort may be made to evaluate the extent of such resources because of their high cost.

Data relating to EAR–II and SR are given in Table 4.2. Few countries estimate resources in these categories, and thus the data are incomplete. The bulk (96 percent) of resources classified as

EAR–II are located in the United States and Canada, while Australia, Canada, and South Africa dominate the SR category.

Very little information is available on the uranium resources of the Centrally Planned Economies (CPE). The OECD/IAEA[4] report World Energy Conference (1980) estimates for low-cost RAR of 208,000 tons U_3O_8 for both China[5] and the Soviet Union, 78,000 tons for the German Democratic Republic, 32,000 tons for Czechoslovakia and 26,000 tons for Romania. Current levels of production

TABLE 4.1 Estimated World[1] Resources of Uranium (as of January 1, 1983) (thousand tons U_3O_8)

	Cost Range (<$30/lb U_3O_8)		Cost Range ($30–50/lb U_3O_8)		Totals	
Country	RAR	EAR–I	RAR	EAR–I	RAR	EAR-I
Algeria[2]	34	—	—	—	34	—
Argentina	24	9	6	—	30	9
Australia	408	480	29	33	437	513
Brazil	212	120	—	—	212	120
Canada	229	235	12	62	241	297
France	73	35	15	8	88	43
Gabon	24	2	6	11	30	13
India	41	6	14	19	55	25
Namibia[2]	155	39	21	30	176	69
Niger[3]	208	69	—	—	208	69
South Africa	248	129	159	62	407	191
Sweden	3	0	48	56	51	56
United States	171	40	359	68	530	108
Other[4]	78	24	78	51	156	75
total	1,908	1,188	747	400	2,655	1,588
total (adjusted)[5]	1,853	1,151	747	396	2,600	1,547

[1] World resources exclude those in China, Eastern Europe, and the USSR.
[2] As of January 1, 1981.
[3] As of January 1, 1977.
[4] Austria, Republic of Cameroun, Central African Republic, Chile, Denmark, Egypt, Finland, Federal Republic of Germany, Greece, Italy, Japan, Republic of Korea, Mexico, Peru, Portugal, Somalia, Spain, Turkey, Zaire.
[5] Adjusted to account for mining and milling losses not incorporated in certain estimates.

Source: OECD Nuclear Energy Agency and the International Atomic Energy Agency, *Uranium Resources, Production and Demand,* OECD, Paris, December, 1983.

TABLE 4.2 Estimated Additional Resources (Category II) and Speculative Resources (as of January 1, 1983) (thousand tons U_3O_8)

Country	Cost Range (<$30/lb U_3O_8) EAR-II	Cost Range ($30–50/lb U_3O_8) EAR-II	Total EAR-II	Cost Range (<$50/lb U_3O_8) SR
Argentina	5	12	17	478
Australia	—	—	—	3,380–5,070
Canada	233	133	366	1,560–1,820
Denmark	—	—	—	13
France	0	16	16	—
Gabon	0	2	2	—
Germany (FRG)	3	3	6	20
Greece	0	8	8	8
Italy	—	—	—	13
Portugal	2	0	2	9
South Africa	—	—	—	1,756
United Kingdom	0	3	3	3
United States	612	440	1,052	898
total	855	617	1,472	8,138–10,088

Source: OECD Nuclear Energy Agency and the International Atomic Energy Agency, *Uranium Resources, Production and Demand,* OECD, Paris, December 1983.

for the Centrally Planned Economies are estimated to be in the range 26,000–32,000 tons U_3O_8 per annum.

SPECULATIVE RESOURCES

In 1976, the OECD and the IAEA established the International Uranium Resource Evaluation Project (IUREP) to evaluate the world's uranium potential and to identify regions in which exploration was likely to be successful. For each of 185 countries an evaluation was made of potential uranium resources, in addition to those already categorized as RAR and EAR, based on the geological favorability for the existence of uranium deposits that could be exploited at costs less than $130/kg U ($50/lb U_3O_8). Given the highly subjective nature of these judgments, and the uncertainties associated with them, these "speculative" resources are reported as broad tonnage ranges on a continent-by-continent basis (see Table 4.3).

Even if these speculative resources exist there is no guarantee that they will be discovered or, if discovered, that they can be made available. These figures, therefore, should be regarded only as a guide for establishing priorities for exploration and evaluation efforts extending into the twenty-first century.

ADDITIONAL SOURCES OF URANIUM SUPPLY

Uranium is widely distributed throughout the earth's surface in approximately the same abundance as tin or molybdenum. While economic resources of uranium are contained in many kinds of deposits, the bulk of the world's known resources occur in the six different types summarized in Table 4.4. In addition to these resources costed at less than \$50/lb U_3O_8, there exist bountiful supplies of higher cost, generally lower-grade, resources. Some of these are extensions of conventional deposits included in the above cost categories, while others may cost in excess of \$50/lb U_3O_8 because of technical or economic factors. Very large amounts of uranium are known to be distributed at very low grade in several areas. In some cases the uranium can be recovered at reasonable

TABLE 4.3 Speculative Resources (million tons U_3O_8)

Continent	Number of Countries	SR (most likely range)
Africa	51	3.4–4.6
America (North)	3	2.7–3.1
America (South and Central)	41	1.3–1.7
Asia and Far East	41	0.7–1.0
Australia and Oceania	18	3.9–4.6
Western Europe	22	0.5–0.8
total WOCA	176	12.5–15.8
		Estimated Total Potential
Eastern Europe, USSR, China	9	6.8–8.5

Source: OECD Nuclear Energy Agency and the International Atomic Energy Agency, *Uranium Resources, Production and Demand,* OECD, Paris, December 1983.

TABLE 4.4 World Uranium Resources by Deposit Type

Category	Average Deposit Grade (% U_3O_8)	Deposit Size Range (tons U_3O_8)	RAR (tons U_3O_8) (<\$50/lb U_3O_8)	EAR (tons U_3O_8) (<\$50/lb U_3O_8)	Principal Locations
Sandstone	0.05–0.3	up to 50,000	1,368,000	1,633,000	Niger, U.S. (Colorado Plateau and Wyoming Basins)
Quartz-pebble conglomerate	0.01–0.15	10,000–200,000	572,000	690,000	Canada (Elliot Lake) South Africa (Witswatersrand)
Proterozoic unconformity-related	0.3–2.5 (and higher)	10,000–250,000	383,000	247,000	Australia (Alligator Rivers area), Canada (Cluff and Key Lakes)
Igneous and metamorphic rocks	0.05–0.15	up to 150,000	269,000	195,000	Namibia (Rossing) France (Lodeve)
Vein	0.1–2.5	up to 30,000	264,000	405,000	Canada (Bear Lake) France (Bessines) Zaire (Shinkolobwe)[2]
Black Shale	0.03–0.04 (Sweden) 0.003–0.005 (United States)	1 million 5–10 million	397,000	—	Sweden (Billingen) U.S. (Tennessee)
Other			131,000	26,000	
total[1]			3,384,000	3,196,000	

[1] Totals differ from those given in Table 4.1 because of different dates of data collection.
[2] Mined out.

Source: Adapted from *Fuel and Heavy Water Availability*, Report of the International Nuclear Fuel Cycle Evaluation. Working Group 1. International Atomic Energy Agency, Vienna, 1980.

cost as a by-product or co-product of another mineral, or by large-scale recovery techniques, while in other cases recovery would not be a viable commercial proposition given current technology. Low-grade uranium resources in conventional-type deposits were considered to be those potentially minable deposits that currently have costs greater than \$50/lb U_3O_8 down to a lower grade limit of 50 parts per million (ppm) U_3O_8. Some examples of low-grade sources of uranium are now considered.

Mill Tailings. Some early uranium mills had low recovery efficiencies. With improved technology there is the opportunity to reprocess these tailings to recover an additional amount of uranium. Some recovery work has been undertaken in the United States, but the quantities involved are not large. About 66,000 tons U_3O_8 contained in the tailings of gold mines in South Africa are included in the estimates of RAR and EAR recoverable at less than \$50/lb U_3O_8. A further 10,000 tons U_3O_8 in tailings are estimated as recoverable at higher costs.

Phosphate Rock. Almost all marine phosphates currently used to produce fertilizer are uraniferous, although the average grade is very low (about 0.01 percent). The uranium can be removed as a by-product of phosphoric acid production, and in 1982 seven U.S. plants, with an annual nominal capacity of 800–1,200 tons U_3O_8, were recovering uranium in this manner. A number of other countries are considering similar ventures.

By-product from Copper Mining. Porphyry copper deposits contain uranium which can be recovered on a small scale from copper leach solutions. This is being accomplished in South Africa and the United States but on a very small scale.

Marine Black Shales. In general, marine black shales contain only low-grade concentrations of uranium. Occasionally, as at the Ranstad deposit near Billingen, Sweden, and at Chattanooga (Tennessee), in the United States, a higher grade deposit may appear. The Swedish deposit is estimated to contain 45,000 tons of recoverable U_3O_8 at a cost of \$30–50/lb U_3O_8 and is included in the RAR for Sweden, while a further 52,000 tons is included in EAR. The bulk of uranium that could be recovered from this deposit, however, falls in the $50–100/lb cost category. The Chattanooga deposit is considerably lower grade, but uranium could be recovered with oil at an estimated (1978) cost of $50/lb. The environmental consequences of mining such large bodies of shale, however, have prevented any mining of the Ranstad deposit.

Coal and Lignites. Although most coals contain very little uranium, certain lignites (a variety of brown coal) contain higher concentrations of uranium. These uraniferous coals have high ash contents and much lower heating values than nonuranium bearing coals, and thus are not suitable as fuel. Uraniferous coals are known to occur in Canada, South Africa, Spain, and the United States, but feasibility studies have shown that recovery costs are well above $50/lb U_3O_8.

Monazite. Uranium is a minor constituent in the mineral monazite, which is mined for its thorium and rare earth content. Current world monazite production is mainly from Australia, Brazil, India, and Malaysia, but none of its uranium content is recovered.

Igneous Rocks. Low-grade granite is a possible source of uranium although, in addition to the very high cost of recovery, environmental considerations would undoubtedly prevent the large-scale stripping operations that would be required to enable its extraction.

Seawater. The world's oceans contain a virtually inexhaustible source of uranium. The concentration, however, is extremely low (about 0.003 ppm), and enormous quantities of water would have to be treated to obtain useful quantities of uranium. With current technology, the cost (about $300/lb U_3O_8 in 1980 dollars) would be prohibitive, although a pilot plant was scheduled to commence operation at Seto Island (Japan) in 1984. If the concept proves feasible, a semicommercial plant is planned by 1990.

Enrichment Plant Tails. With a 0.20 percent tails assay at the enrichment plants, about 23 percent of the uranium-235 from the natural uranium remains in the plant tails. Recovery would require substantial improvements in enrichment technology (thus reducing the cost of SWUs) or a substantial increase in the price of uranium. Under such circumstances, however, the accumulated enrichment tailings could develop into a significant source of uranium supply.

URANIUM EXPLORATION

We noted earlier in this chapter that uranium is widely diffused throughout the earth's crust and oceans, but that deposits that are sufficiently rich to be economically exploitable are quite rare. Uranium has an average crustal abundance of 4 ppm, yet a minimum concentration of about 700 ppm is required to allow commercial exploitation under current technological and economic conditions. Although exploitable reserves are relatively rare, fortunately they appear not to be randomly distributed across the globe but are frequently discovered on the world's great "shields." Mineral exploration is an expensive, risky, and time-consuming business, but uranium exploration has the advantage that it can be conducted from the air using radiometric techniques.

For the majority of developed nations, national geological surveys provide a basis for mineral exploration, but this facility is typically absent in most developing nations; as are finance, qualified manpower, and the administrative and technological infrastructure required to support exploration effort. In recent years, major exploration programs in developing countries have concentrated in Algeria, Argentina, Brazil, Gabon, India, Namibia, and Niger (prior to the 1979 Islamic Revolution, Iran also was involved), although IUREP "orientation" missions have covered many more. These missions attempt to collect information and provide advice for better evaluation of a country's uranium potential. For many developing nations, the technological and financial requirements of exploration programs have been provided by the major uranium-consuming (developed) nations,[6] although ex-colonial links (e.g., France with Gabon and Niger) have also been a factor. For others, who wish to control the destiny of their own uranium exploration programs, projects are generally government-sponsored and frequently appear to attract insufficient interest or support. A notable exception is Brazil, where the government has engaged in an extensive program of uranium exploration with the aim of becoming self-sufficient in the fuel requirements for its proposed nuclear-power program.

Worldwide, exploration expenditure increased dramatically during the second half of the 1970s (Tables 4.5 and 4.6), stimulated by the rapid rise in uranium prices and the threat of an impending world energy shortage. Some industrialized nations with inadequate domestic energy supplies, principally Japan and several Western European countries (notably France), were compelled to

TABLE 4.5 Uranium-Exploration Expenditure in Countries Listed ($ million)

Host Country	pre-1977	1977	1978	1979	1980	1981	1982	1983[1]
Argentina	n.a.	4.8	7.2	5.6	7.4	5.2	4.1	n.a.
Australia	101.0	19.0	29.0	33.0	45.0	43.0	n.a.	n.a.
Brazil	77.2	25.6	29.3	12.5	8.6	18.8	17.9	4.3
Canada	92.7	67.4	78.9	110.5	109.5	85.2	57.5	34.5
France	139.6	18.2	26.2	61.2	89.5	70.5	55.7	50.7
Gabon	33.7	4.8	8.5	9.6	8.6	8.3	7.7	5.9
Germany (FRG)	15.7	9.3	11.5	14.4	12.7	13.3	7.4	6.0
India	53.0	4.4	4.9	7.7	n.a.	n.a.	n.a.	n.a.
Namibia	n.a.	2.9	3.7	1.6	2.8	2.5	2.0	n.a.
South Africa	n.a.	12.4	24.6	24.1	22.1	19.2	6.8	5.3
Spain	33.6	6.8	8.0	12.0	11.0	16.0	16.0	16.0
Sweden	12.7	5.4	5.9	5.0	5.5	5.1	3.4	2.4
United States	729.8[2]	293.5	371.5	385.6	332.6	180.3	97.8	56.7
Other WOCA	70.3	11.6	14.4	26.8	27.2	13.8	10.3	8.1
total	1,359.3	486.1	623.6	709.6	682.5	481.2	286.6	189.9
total (constant 1983 dollars)		757.7	901.3	913.4	768.7	496.4	289.9	189.9

[1] Planned.
[2] 1966–76 only.

Source: OECD Nuclear Energy Agency and the International Atomic Energy Agency, *Uranium Resources, Production and Demand*, OECD, Paris, December 1983.

TABLE 4.6 Uranium-Exploration Expenditures Abroad by Countries Listed ($ million)

Funding Country	pre-1973	1973	1974	1975	1976	1977	1978	1979	1980	1981	1982	1983[1]
Belgium	0	0	0	0	0	0	1.1	1.2	0.9	0.5	0.5	0.3
France	66.8	14.9	19.0	22.7	31.9	32.0	36.9	52.3	68.2	66.2	40.4	26.5
Germany (FRG)	41.0	6.0	25.0	14.0	21.0	21.0	28.0	30.0	30.0	26.0	26.0	26.0
Japan	9.5	1.2	1.4	2.9	8.4	24.0	24.4	24.5	29.3	30.2	24.2	24.2
Korea (South)	n.a.	n.a.	n.a.	n.a.	n.a.	n.a.	1.0	1.9	5.7	4.0	3.7	3.3
Spain	0	0	0	4.1	1.3	1.4	2.5	4.0	2.0	2.0	3.0	0.2
Switzerland	0	0	1.5	1.5	0.8	2.6	2.2	2.9	3.2	2.8	1.7	n.a.
United States	17.0	3.0	5.0	5.1	18.8	31.2	35.9	43.2	39.0	35.3	14.9	6.1
total[2]	134.3	25.1	51.9	50.3	82.2	112.2	132.1	160.2	178.4	166.9	114.5	86.6
total (constant 1983 dollars)						174.9	190.9	180.4	200.9	172.2	115.8	86.6

[1] Planned.
[2] May not equal the sum of the expenditure figures due to rounding.

Source: OECD Nuclear Energy Agency and the International Atomic Energy Agency, *Uranium Resources, Production and Demand*, OECD, Paris, various issues.

increase substantially their financial interest in overseas exploration ventures. The notable exception to this trend was Australia, where the development of the newly discovered deposits in the Northern Territory was frozen pending a federal government inquiry into the economic, social, and environmental costs and benefits of uranium mining. Exploration expenditure on uranium actually declined in Australia following the 1973 OPEC crisis, and it was not until 1977 that it passed (in real terms) its 1973 level.[7]

The data given in Tables 4.5 and 4.6 are expressed in terms of both current (or nominal) and constant (1983) dollars. The economies of both developed and developing nations were characte-

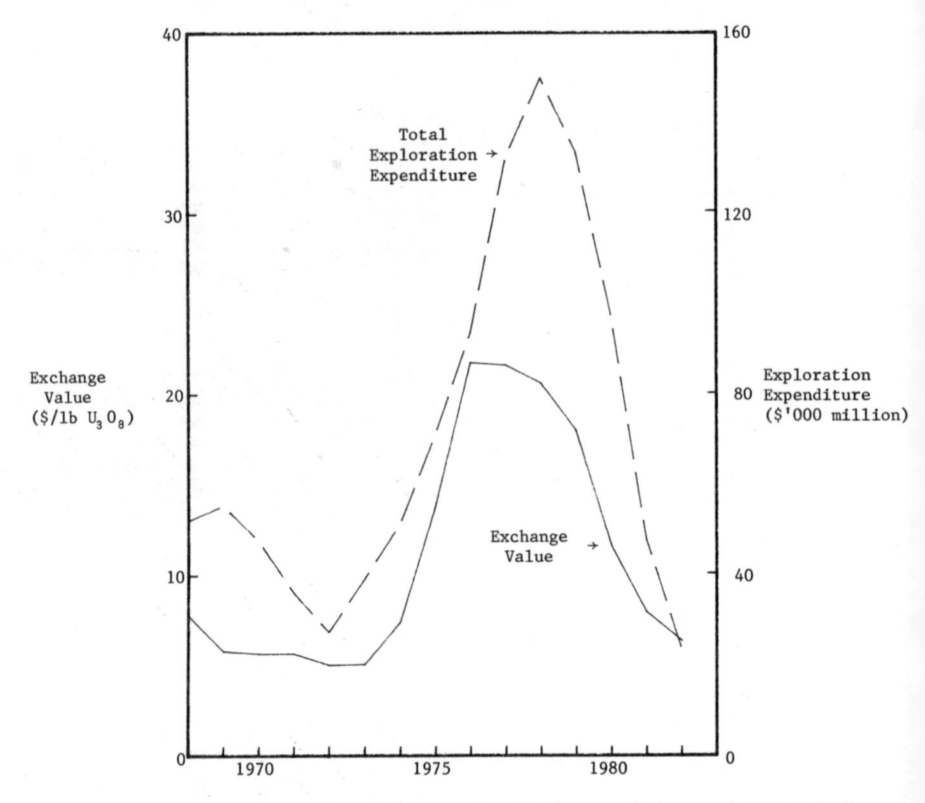

FIGURE 4.2 The Relationship between the Exchange Value and Total U.S. Exploration Expenditure.

Source: Total exploration expenditure: *Statistical Data of the Uranium Industry,* U.S. Department of Energy, various issues. Exchange value: *NUEXCO* reports.

Both series were deflated by the U.S. Producer Price Index for Industrial Commodities (1967 = 100).

rized by relatively high (by post-1945 standards) rates of inflation during the second half of the 1970s, and the current dollar figures therefore give an exaggerated impression of the real increase in exploration. In terms of constant dollar values, the peak year for world exploration expenditure was 1979, after which it dropped dramatically. To illustrate this point, Figure 4.2 shows the relationship between the spot price of uranium (i.e., NUEXCO's exchange value) and domestic U.S. exploration expenditure since 1969,[8] both series being expressed in terms of 1967 dollars. Between 1972 and 1978, U.S. exploration expenditure increased by about 450 percent in real terms (a little over 900 percent in nominal terms) in response to the rapid rise in the exchange value. The post-1978 decline in both variables has been equally as dramatic. The simple correlation coefficient between the exchange value and the following year's expenditure on exploration is 0.94, which reflects not only the high degree of association between these two variables but also the speed with which exploration can respond to price stimuli.

NOTES

1. See Chapter 1 for a listing of these reports.
2. Prior to the 1983 OECD/IAEA report, Estimated Additional Resources (EAR) was a single category. It referred to uranium in addition to RAR that is expected to occur, mostly on the basis of direct geological evidence in: extension of well-explored deposits; little-explored deposits; and undiscovered deposits believed to exist along a well-defined geological trend with known deposits. In general terms, the "discovered" part of the resources in EAR is reported in EAR-I in the 1983 report, while the undiscovered part is in EAR-II.
3. OECD/IAEA (1983), pp. 9–11.
4. OECD/IAEA (1982).
5. More recently, Geddes (1983) has reported China's reserves to be an estimated 800,000 tons uranium, with total resources (including speculative resources) of 3 million tons uranium. These figures rank China's uranium resources as among the largest of any country in the world.
6. Zorn (1982) discusses Third World involvement in uranium and other industries and the disadvantageous influences which may arise from foreign involvement and/or control of the domestic industry.
7. Details of events that constrained the growth of the Australian uranium industry during the late-1970s are given in Chapter 6.
8. The U.S. exploration expenditure data used in constructing Figure 4.2 differ from those presented in Table 4.5. Two possible explanations for the former being consistently about 20 percent below the latter are: (a) the DOE data used

in Figure 4.2 are based upon an annual survey of uranium-exploration companies, and less-than-adequate responses may have caused the true figure to be underestimated; (b) there may be difficulties associated with allocating exploration expenditure (especially in the early stages of exploration) specifically to uranium. Typically, uranium is found in association with other minerals (e.g., lead, copper, gold) and it may be difficult to allocate to uranium its share of exploration expenditure when the mining company may wish to extract uranium as a co-product or by-product.

5

The Economics of Uranium Supply

THE ECONOMICS OF AN EXHAUSTIBLE RESOURCE

Since uranium, in common with all other minerals, is an exhaustible resource, its extraction and use today involves a cost (in addition to normal economic costs) to the producer in terms of the sacrifice of future profits. The present value of these future profits would determine whether it was more profitable to extract the resource today or to hoard it until a later date. It follows that the optimal level of extraction of an exhaustible resource in any one time period will be determined in such a way that the sum of the discounted marginal profits over all future time periods is maximized.[1]

In a perfectly competitive market (i.e., a large number of buyers and sellers acting independently, with perfect knowledge, and trading in a homogenous product), elementary economic theory dictates that the profit-maximizing output for an individual firm will occur when marginal cost and marginal revenue (and hence price) are equal. When considering an exhaustible resource, however, an additional cost (generally known as "user cost") is incurred due to the nonreplenishable nature of the product. The user cost is the opportunity cost of holding the resource *in situ*; i.e., it is the present value of all future sacrifices associated with the use of an *in situ* resource. If this opportunity cost is less than the current value of the resource then it would benefit the producer to sell the resource and invest the money in income-earning assets. If the opportunity cost is greater than the current market value of the resource, however, then clearly it would benefit the producer to leave the resource *in situ*.

The profit-maximizing output for a firm engaged in the extraction of a nonreplenishable resource, therefore, is where marginal revenue is equal to marginal cost *plus* the user cost of the marginal unit being extracted at any point in time. The latter cost is generally interpreted as a royalty (or a scarcity rent); i.e., it is the cost of holding the resource *in situ*.

If extraction costs are assumed to be negligible, the implication of the concept of user cost is that, in equilibrium, the scarcity rent (and hence the future price of the resource) must increase at the rate of interest (r). In the diagram below, S_0 denotes the market supply curve (which is simply the sum of the supply curves of the individual producers) in the current period, and D denotes the aggregate demand curve. In the next period, the supply curve, S_1, must lie to the left of S_0 because of depletion of the resource. Production decreases from q_0 to q_1, while the price rises from p_0 to p_1, where $p_1 = p_0(1 + r)$. The higher the rate of interest, the greater will be the current rate of depletion of the resource and hence the farther S_1 will be to the left of S_0.

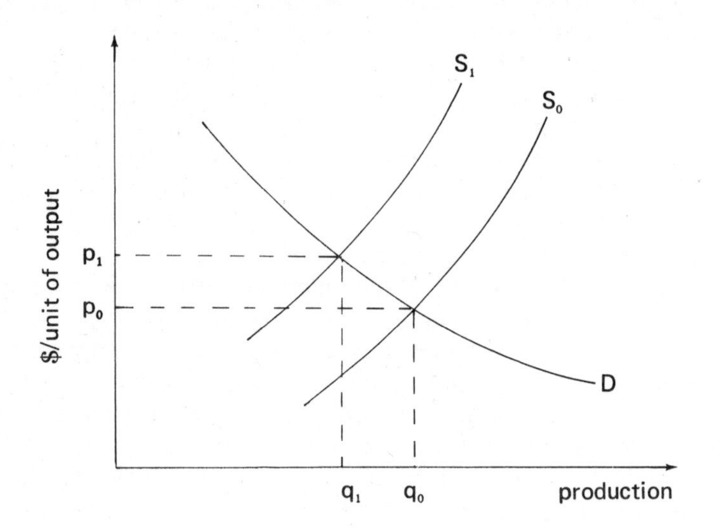

If extraction costs are now considered to be nonzero, but constant, the rate of price increase will equal the rate of interest multiplied by the share of scarcity rent in price. Again, the market price of the resource rises over time.

This analysis can be extended to allow for variations in production costs between producers. The marginal resource will receive a price equal to the user cost associated with its extraction (i.e., scarcity rent), while lower cost deposits will command a premium, known as Ricardian rent, equal to the production cost saving on such resources.

The analysis can be extended further by allowing for additions to the endowment of the resource by investment in exploration. When this occurs the total stock (i.e., reserves) of the resource is increased and consequently price and royalties are reduced. Thus as the stock of a resource is depleted, increased prices provide an automatic incentive for conservation, exploration, recycling, and technological developments aimed at the provision of substitute commodities.

PRACTICAL CONSIDERATIONS

Clearly the above analysis is provided within a very simplistic framework. A number of factors will now be considered which have the effect of modifying this theoretical approach to allow for the impact of more practical considerations. While all of the factors considered below can be regarded as general in nature, we will discuss them in the specific context of the uranium industry.

Perfect Knowledge

As planning horizons stretch further into the future, both producers and consumers of uranium face greater uncertainty with regard to the impact of current decision making. For example, the lead time between discovery of a new uranium deposit and the commencement of production on a commercial scale generally ranges (depending on the country where the deposit is located) from five to ten years. While a whole host of geological and economic factors may shorten or lengthen the lead time for a single project, during the late 1970s and early 1980s a greater social awareness of environmental issues and the rights of (generally) underprivileged native people have combined to extend lead times (considerably in Australia and, to a lesser extent, Canada). Thus, for example, when Pancontinental's Jabiluka deposit was discovered in the Northern Territory of Australia in 1972 it was envisaged that full production would be achieved by the end of 1978. The current (1985) Labor government, however, is opposed to expanding the domestic uranium mining industry. As a consequence, this deposit is unlikely to become operational before 1990. The uncertainties inherent in such a situation, combined with uncertainties surrounding the future of nuclear energy in many nations, serve both to delay mining projects and increase the risk for investors in such projects.

The delays mentioned above have now become an entrenched part of the uranium mining industry and, while knowledge of them may not be "perfect," the owners of all new uranium ventures will expect to have to confront a great number of environmental and social conditions as well as, in many countries, a certain amount of public hostility before mining can commence.

Exogenously determined shocks (often initiated by domestic governments) are a different matter. Frequently these have had a major unforeseeable impact upon prices and consequently upon both current and future production and exploration activity. For example, the U.S. embargo on the enrichment of foreign uranium for domestic consumption decimated the Canadian uranium industry virtually overnight.

Another example is the introduction in 1973 of the Long Term Fixed Commitment Contract for enrichment services provided by the United States (the world's sole provider of such services at the time) which, combined with the OPEC oil embargo and subsequent oil price rises, initiated the explosion in the exchange value in 1974 and 1975. The momentum was sustained by the Westinghouse failure in September 1975. In retrospect, of course, the signs of many of these exogenous shocks were in evidence to "experts" many years before their impact. The fact remains, however, that in general the industry was either ignorant or reluctant to act and was unable to meet the changing conditions due to the long lags and lead times involved in adjusting uranium production to changes in the conditions of demand (and hence prices).

Physical Availability

Physical availability is not necessarily a good measure of economic abundance. Many technical factors may limit or prevent the mining of uranium by underground cutting, while open-cut mining in many areas may be environmentally unacceptable. Chapter 4 considered a number of methods of obtaining vast supplies of uranium, but for either economic (e.g., the extraction of uranium from seawater) or environmental (e.g., the Ranstad deposit in

Sweden) reasons they could not be categorized as available "resources."

Private vs. Social/Public Costs and Benefits

For the individual producer, gross profits can be maximized by maximizing the discounted flow of gross profits over the life of the project. If free entry to the industry is assumed, then a producer may enter or leave a project according to its relative profitability.

For a nation as a whole, however, the availability of resource deposits may be restricted by government policy aimed at preserving some deposits for future generations or at ensuring that adequate domestic supplies are available for the foreseeable future.

For projects which are operational, both direct and indirect social costs and benefits must be considered in the same monetary terms as those of a private nature. For example, many uranium mining ventures are in remote locations with poor infrastructure and high rates of local unemployment. The social benefits which may arise from the location of a mine in such an area are improved infrastructure (itself a private cost) and a more pleasant social environment due to the reduction of unemployment and a growing local economy. High among social costs is environmental damage (especially with open-cut mining) which, although a private cost since rehabilitation of the natural landscape on the completion of mining is a condition for most new mines in developed nations, will necessarily lead to long-term disruption of the natural environment. The prospect of lower levels of local unemployment, however, may be reduced by the necessary hiring of nonlocal skilled labor. Also, typically, little of the money generated by mining projects in isolated areas adds to local prosperity since the principal items of expenditure are invariably made in the more industrialized parts of the nation. It was with these latter points in mind that the Saskatchewan government insisted that substantial financial benefits in the form of aid were to flow to the local inhabitants from the establishment of the Cluff Lake and Key Lake mining projects in the northern part of the province.

Recycling

By extending the stock of a resource through recycling, its price would be lowered accordingly. Reserves and resources of uranium can be extended by recycling spent products at various stages of the fuel cycle, and this topic was considered in detail in Chapter 2. In particular, spent fuel reprocessing can extend resources of both uranium and plutonium, although the economics of recycling spent fuel may argue against such action in the immediate future. Since spent nuclear fuel passes through relatively few hands, if recycling is found to be an economically viable proposition sometime in the future, it would be a relatively simple matter to recover the vast majority of it. By way of contrast, consider the enormous number of outlets for aluminum products and the problems associated with collecting material suitable for recycling. Nevertheless, the secondary aluminum industry thrived as the cost of power rose during the 1970s and early 1980s, since production of secondary aluminum only requires one tenth of the power used in primary aluminum production.

Environmental Considerations

To some extent environmental considerations were covered by the discussion on private versus social costs. The imposition of conditions requiring revegetation of mining areas or, equivalently, a tax to cover environmental damage will raise the cost of extracting the resource. In some cases environmental damage may be considered so undesirable that development of the project itself, irrespective of its potential value to the nation, would be prohibited by the relevant authority. For example, the Koongarra project in Australia's Northern Territory lies on the edge of the Kakadu National Park, and the Ranger Uranium Inquiry recommended that development of the project be prohibited to avoid irreparable damage to the sacred Aboriginal sites contained within the park.

Reserves and Resources

As with any exhaustible resource, the expected trend in the price of uranium depends essentially on the reserves of uranium that are expected to be available at various costs of production. Uranium reserves are reported by the OECD using a "forward-cost" concept; i.e., the reserves that could be mined at forward costs of (a) less than \$80/kg U (\$30/lb U_3O_8) and (b) less than \$130/kg U (\$50/lb U_3O_8). The U.S. DOE uses four forward-cost categories (less than \$15/lb U_3O_8, less than \$30/lb U_3O_8, less than \$50/lb U_3O_8, and less than $100/lb U_3O_8) for reporting domestic resources of uranium. We noted in Chapter 4, however, that the forward-cost concept should not be confused with the price of uranium as, in general, the latter would need to be substantially higher than the former to encourage extraction of these resources. As uranium reserves are depleted, we have already noted that increased prices provide an automatic incentive for exploration. For strategic reasons, however, countries with nuclear-power programs may decide that domestic exploration in excess of the level determined by market forces is desirable in order to minimize their reliance on foreign supplies of uranium.

Technological Change

Technological change is probably the most important long-term factor in determining future levels of both production and consumption of any commodity. Not only can technological progress extend the life of a commodity but, given a sufficient incentive, it can account for the development of substitutes. In the case of uranium production, laser enrichment could extend considerably the active life of the finite amount of uranium that the world has at its disposal, whereas the widespread introduction of the FBR would place an upper boundary on consumption requirements of "new" uranium. Substitutes in the form of solar or fussion energy (or both) could mean, ultimately, the demise of the nuclear fission industry (and hence the uranium mining industry unless other uses for uranium are found on a comparable scale).

Industry Organization

Modern-day extractive industries are frequently dominated by a small number of producers, who attempt to control between them various aspects of the market. Such organizations often have a range of interests and goals, but their basic concern is to ensure adequate recompense for the extraction of their nonreplenishable resources. More sinister policy objectives may of course be pursued, but generally this form of association can be viewed as an attempt to make the future less uncertain.

The most influential and highly publicized producer organization (or cartel) in recent years has been OPEC, but many other such organizations have experienced (at least partial) success in manipulating market conditions to reap substantial returns for their members. The International Bauxite Association, the International Tin Producers Association, and the De Beers diamond group are just a few that have enjoyed varying degrees of success, but their impact on the world economy has been of minor significance compared with that of OPEC. Even uranium producers had a short-lived attempt at cartelization which, in retrospect, would generally be regarded as disastrous.

URANIUM SUPPLY

Any analysis of the economics of uranium supply must necessarily consider a number of interdependent economic, technical, and sociopolitical variables, which combine to determine the actual level of mine production. Time, however, is a critical factor in the process, and thus our analysis of supply must consider variables whose impact on the market extends from the immediate to the long term.

Ultimately, the potential (long-term) supply of uranium is a function of total reserves and resources, which is itself dependent upon the level of exploration activity. If reserves are low relative to a producer's current and future expected levels of production, then expenditure on exploration can be justified. If reserves are adequate for projected requirements over a relatively long period, however, the discounted cost of investing in exploration today to provide reserves for a distant future could be considerable. Since exploration is generally undertaken many years before the first sales of yellowcake, price expectations will play a major role in exploration effort. Typically the lead time between initial exploration and eventual discovery and mining of a deposit can take up to 15 years. While exploration can be expanded and contracted fairly readily, results take considerably longer to achieve.

Mining companies cannot vary their production to a significant degree once the capital works of the project have been established and mining has begun. Large variations in output would be prohibitively expensive and, in the case of temporary or permanent closing of the mine, would result in the loss of a skilled work force and occasionally the demise of entire towns. Generally production is planned in advance to cover the mine's estimated working life, and major deviations from these plans would only occur in exceptional circumstances.

While a decrease in production produces fewer technical problems than increasing production, any substantial drop in production will result in higher unit costs. The fixed costs of mining must be met irrespective of the amount produced, and, especially in the case of underground mines, these can be extremely high. Depending on market expectations, it may be less costly to stockpile excess production rather than close the mine.

Short-run Supply Flexibility

As the price of uranium rises (in real terms), a corresponding short-run increase in mine production will be desired if producers are attempting to maximize their profits. As the design capacity of individual mines and mills is reached, however, a number of potential bottlenecks may arise.

Ore Grade. To some extent it may be possible to raise the grade of ore delivered to the mill. While this may not represent too much of a problem in open-cut mines, it is probably a less feasible proposition in underground mines. This can only be regarded as a short-term remedy, however, as a prolonged period of selective high-ore-grade mining could adversely affect long-term mining schedules.

Alternatively if a company is concerned with maximizing the life of its mine, rather than profit, it may actually reduce its yellowcake production by mining ore of a lower grade, which the price rise has now rendered economic.

Utilization Rates. Mine and mill design capacities are generally expressed in terms of the ore throughput per day. Mill output of yellowcake, however, will vary according to the grade of ore and its amenability to treatment (i.e., its hardness and abrasiveness). In the short run it would be possible to extend the capacity of the mill beyond its design capacity by eliminating routine maintenance and shutdowns. This may allow for additional mill capacity to be constructed (about two years), but its long-run effect is likely to be a rise in costs and unplanned maintenance.

Recovery Rates. At several points in the nuclear fuel cycle, there is scope for an improvement in recovery rates, subject to operating costs. If the price of uranium rises we have already noted that the effective supply will be increased via a lower enrichment tails assay; i.e., the recovery rate at the enrichment plant would be higher. There is similar scope for increased recovery rates at the milling stage.

By-product Uranium. When uranium prices are low, the processing of uranium that is mined as a by-product of another mineral may not be justifiable. If prices rise, however, processing of the previously mined "wastes" may become economical. This is particularly relevant for South Africa, where uranium is mined as a by-product of gold and can simply be stockpiled if prices do not warrant its processing at the time of extraction.

Labor Shortages. Since the majority of mineral industries experience all phases of the business cycle more or less simultaneously, their demand for skilled labor tends to rise and fall together. Thus in times of rapid economic growth skilled labor tends to be at a premium at the mine sites. This is particularly noticeable in South Africa where nonwhites are forbidden to rise to skilled labor status and a chronic shortage of skilled (white) operatives arises in times of high mineral demand.

The Long-run Supply Function

While the short-run supply function reflects the relationship between price and output with a given (i.e., constant) level of mine and mill capacity, the ultimate supply response to given price levels is reflected by the long-run supply function. The impact on producers of an increase in the price of uranium will depend largely upon the current state of the industry and expected future price movements. If mines and mills are working at close to capacity and inventories are relatively low, then producers have relatively little scope for expansion in the short run due to the constraints of mine and mill capacity. This situation is illustrated in Figure 5.1, where the short-run supply curve is denoted by S_0 and the current price is denoted by P_1. At this price, however, the amount of uranium that it is profitable for producers to supply in the long run is Q_1. If the current price (P_1) is not perceived as a temporary phenomenon, then entry into the industry and expansion of existing production facilities will occur and the short-run demand curve would ultimately be shifted to a new equilibrium position represented by

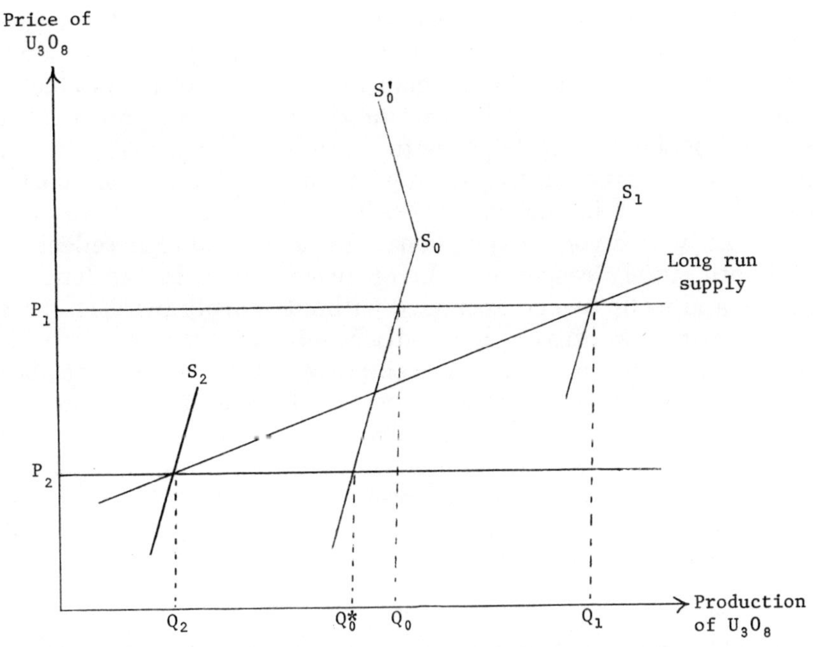

FIGURE 5.1 The Relationship between Short-run and Long-run Supply Curves.

S_1. Until this position is reached, however, the short-run price will be above the long-run price. Similarly, if the price falls to P_2 then output in the short run will only fall from Q_0 to Q_0* whereas, given time, closing of some production facilities and contraction of others will shift the short-run supply curve to a new equilibrium position denoted by S_2, at which point production will have fallen to Q_2.

The fact that supply is relatively fixed (i.e., inelastic) in the short run with respect to price movements is reflected by the slope of the short-run supply curve, which indicates that a relatively large price increase is required to induce the quantity supplied to rise by even a small amount. Indeed, it is possible for the short-run supply curve to bend backward (as represented by the line S'_0S_0 in the diagram); i.e., for production actually to fall as the result of a price increase. If, as prices rise sharply above expected levels, producers process waste heaps and stockpiles of low-grade ore and mine lower-grade ores more extensively, then output of U_3O_8 may actually decline in the short run due to the decline in ore grade, although mill throughput will, in fact, increase.

The supply response of producers to different price levels takes place at several related levels of decision making. The decisions to explore for new deposits, develop reserves for future production, and to mine, mill, and sell uranium from developed properties are all part of the supply response to current and expected prices and give rise to lags in the timing of the production response to expectations of increased prices. The long-run supply response, however, will involve expansion at all levels, since it reflects the ultimate supply response to given price levels. In the long run, mines and mills can be expanded without sharply increasing costs and existing reserves can generally be mined more intensively. The major constraint, and hence the potential source of substantial cost increases, is the availability of reserves to replace those currently being mined. In general, this would involve substantial sums being spent on exploration although, in the case of uranium, the potential for the discovery of new, high-grade, deposits is still very high in many countries of the world.

PRICE RESPONSIVENESS OF MINE PRODUCTION

Figure 5.2 illustrates the relationship between NUEXCO's exchange value and Canadian, South African, and U.S. mine production of uranium since 1968. The simple correlation coefficients between the exchange value (yearly average) lagged three years and Canadian and South African mine production are 0.84 and 0.95, respectively. The corresponding price lag for U.S. mine production amounted to between two and three years, with correlation coefficients of 0.94 and 0.85, respectively.

These results reflect not only the high degree of linear association between mine production and the exchange value, but also the long lag structure implicit in the relationship. This latter point can be illustrated by considering the price explosion in the mid-1970s. U.S. mine production did not peak until 1980-81, more than six years after the initial stages of the price rise. By year-end 1980, the exchange value had actually been declining for two years, although in real terms the decline started in 1977.

While South African and U.S. mine production follow similar trends over the entire period, it is worth noting that Canadian mine production does not experience a downturn after 1980. In a climate of rapidly falling prices, and with the gradual removal of the U.S. embargo on the domestic use of imported uranium, U.S. uranium

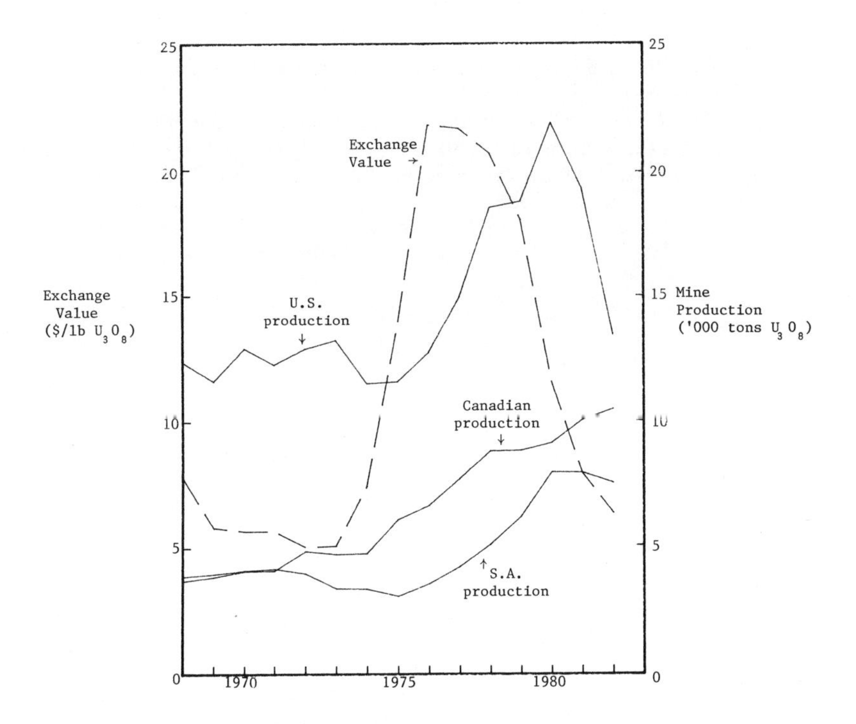

FIGURE 5.2 The Relationship between the Exchange Value and Mine Production in Canada, South Africa, and the United States.
Source: Mine production, from Table 3.1. Exchange value (deflated series, 1967 = 100), from Table 3.2.

production fell rapidly after 1980 as high-cost domestic producers were forced to cut back production or close entirely. The latest OECD/IAEA Red Book forecasts that this trend will continue until the mid-1980s, by which time U.S. production should have stabilized at around 13,000–14,000 tons U_3O_8, rising to about 16,000 tons U_3O_8 by 1990.[2] While South African uranium production has also fallen since 1980, the extent of the decline has been relatively small and is expected to remain so. To a limited degree, the decline in U.S. production is being offset by an expansion of production from the low-cost reserves in Saskatchewan, Canada. A more detailed analysis of the production potential of the world's major uranium producers is given in the next chapter.

NOTES

1. For a more detailed analysis of this topic the reader is referred to Banks (1983) or Howe (1979).
2. OECD/IAEA (1983). These figures must now be considered unrealistic. It is unlikely that U.S. uranium production will rise above 10,000 tons U_3O_8 before 1990 (refer to estimates provided by NUEXCO, *Monthly Report on the Nuclear Fuel Market,* October 1983).

6

World Uranium Production

INTRODUCTION

Currently, just eight countries account for the bulk (about 98 percent) of the Western world's production of uranium, with little possibility of this group being expanded over the next decade. Brazil is the only country with substantial uranium reserves that does not currently possess a significant production capability. There are no plans, however, for expansion of Brazil's production beyond a level capable of satisfying its modest domestic uranium requirements. Current and projected levels of production for these eight countries are given in Tables 6.1 and 6.2. Table 6.1 presents projections of short-term uranium production capabilities made by OECD/IAEA in 1983, while Table 6.2 presents NUEXCO estimates of future production levels which were published in the same year. When comparing these two sets of estimates, the remarkable difference between the two projections for U.S. production through to the early 1990s is immediately apparent.

The projections given in Table 6.1 show the "maximum levels of production that could be practically and realistically achieved under favorable circumstances from the plant and facilities at specific production centers within each country." These projections envisage a very small rate of growth in world production capability to 1990, with a period of stagnation thereafter. The NUEXCO forecasts given in Table 6.2 refer to actual production and not production capabilities. Nevertheless, with the notable exception of the United States, the data are very similar.

TABLE 6.1 Short-Term Uranium Production Capability Projections (thousand tons U_3O_8)

Country	1984	1985	1986	1987	1988	1989	1990	1991	1992	1993	1994	1995
Australia	4.94	4.94	4.94	3.90	3.25	4.29	4.29	5.85	5.85	6.50	6.50	6.50
Canada	13.65	14.95	15.60	15.86	15.86	15.73	15.73	15.60	15.21	14.95	14.30	12.87
France	5.07	5.07	5.07	5.07	5.07	5.07	5.07	5.07	5.07	5.07	5.07	5.07
Gabon	1.95	1.95	1.95	1.95	1.95	1.95	1.95	1.95	1.95	1.95	1.95	1.95
Namibia	5.07	5.07	5.07	5.07	5.07	5.07	5.07	5.07	5.07	5.07	5.07	5.07
Niger	5.20	5.20	5.20	5.20	5.20	5.20	5.20	5.20	5.20	5.20	5.20	5.20
South Africa	7.97	8.22	8.64	8.64	8.49	8.40	8.40	8.40	8.40	8.18	8.18	7.95
United States	13.39	13.52	14.17	14.69	15.21	15.60	15.86	17.29	17.16	17.16	18.20	18.20
Other	1.71	1.74	1.80	1.79	2.48	2.33	2.59	2.58	2.58	2.58	2.30	2.84
total	58.95	60.66	62.44	62.17	62.58	63.64	64.16	67.01	66.49	66.66	66.77	65.65

Source: OECD Nuclear Energy Agency and the International Atomic Energy Agency, *Uranium Resources, Production and Demand,* OECD, Paris, December 1983.

TABLE 6.2 World Uranium Production Forecast: NUEXCO (thousand tons U_3O_8)

Country	1983	1984	1985	1986	1987	1988	1989	1990	1991	1992
Australia	5.25	5.25	5.25	5.25	5.25	4.60	4.00	5.00	6.40	8.80
Canada	10.45	12.45	16.05	15.70	15.95	15.95	15.95	15.95	15.80	15.80
Central Africa[1]	5.60	5.10	4.85	4.85	4.85	4.85	4.85	4.85	4.85	4.85
France	4.00	4.00	4.00	4.00	4.00	4.00	4.00	4.00	4.00	4.00
Namibia	4.90	5.00	4.25	4.10	3.25	3.25	3.25	3.25	3.25	3.25
South Africa	7.80	8.15	8.40	8.50	8.55	8.65	8.75	8.70	8.70	8.50
United States	8.15	6.10	7.65	8.55	8.60	9.10	9.15	10.45	11.25	11.15
Other	1.20	1.40	1.40	1.45	1.45	1.45	1.35	1.20	1.55	1.90
total	47.35	47.45	51.85	52.40	51.90	51.85	51.30	53.40	55.80	58.25

[1] Gabon and Niger combined.

Source: NUEXCO, Monthly Report on the Nuclear Fuel Market, October 1983.

NUEXCO predicts that U.S. production will continue its post-1980 decline (see Table 3.1) throughout the first half of the 1980s but will experience a modest revival after 1990. However, the OECD/IAEA projections for the United States envisage a continuous, but relatively small, expansion of U.S. production capability throughout the period of forecast. As a consequence, the OECD/IAEA projection for 1990 is more than 50 percent higher than the corresponding NUEXCO figure.

While NUEXCO estimates for the other seven countries are also generally lower than the corresponding OECD/IAEA figures, the difference is far less dramatic than for the United States. Only Canada is predicted to experience a significant growth in production during the first half of the 1980s, with the Key Lake deposit reaching full production capacity, and by 1990 is expected to account for about 30 percent of world production. A surge in Australian production is predicted to occur around the turn of the decade and to gather momentum in the early 1990s when the Olympic Dam deposit comes on-stream. For all other countries, however, production is expected to remain fairly static.

In this chapter the current status and prospects for the major uranium-producing nations of the Western world are now considered on a country-by-country basis (in alphabetical order).

AUSTRALIA

Before the Second World War, uranium was a mineral of only minor commercial interest in Australia. Uranium ore had been mined intermittently for the recovery of radium, but little of the uranium content was recovered. The rapid wartime development of nuclear technology for military purposes, however, created a commercial market for uranium, and in 1944 exploration for uranium deposits began in Australia at the request of the U.K. government.

The advent of the Cold War further stimulated the demand for uranium, and hence intensified the worldwide search for uranium deposits, as the world's superpowers embarked on a rapid expansion of their nuclear arsenals. In Australia, private exploration was encouraged by tax-free rewards from the federal government for the discovery of uranium deposits. All marketing was controlled by the federal government, which offered guaranteed prices for uranium ores of various types and grades as well as tax concessions for companies mining and treating uranium ore. Most of the uranium mined in Australia during the period 1954–64 came from the Rum Jungle (Northern Territory) and Mary Kathleen (Queensland) deposits and was produced to meet export contracts with the Combined Development Agency.

While the market for uranium began to decline in the late 1950s, the Australian deposits that were being mined (with the exception of Mary Kathleen) were also near depletion. By 1964 all uranium mining in Australia had ceased, although production of uranium from stockpiled ore continued until 1971. No uranium ore was mined on a commercial scale in Australia from 1964 to 1975.

The expansion of nuclear-power programs in Western nations during the late 1960s and early 1970s promised a prosperous future for the Australian uranium-mining industry. Exploration by private companies started in earnest, and important new deposits in the Northern Territory, at Koongarra, Nabarlek, and Ranger, were announced in 1970. By the middle of 1971, more than 80 companies were engaged in uranium-exploration programs, and during the following year major discoveries were announced at Yeelirree (Western Australia) and Jabiluka (Northern Territory). By now it was apparent that Australia had the reserves to become a major exporter of uranium in the 1980s. In mid-1973 total uranium reserves (recoverable at a cost of less than \$10/lb U_3O_8) were estimated to be 140,000 tons U_3O_8, about 25 percent of the world's known reserves.

Between 1970 and 1972 Australian companies obtained contracts for the export of approximately 11,500 tons U_3O_8 for delivery during the period 1976–86. In early 1973, however, the federal government indicated that approval for all new export contracts would be withheld, pending the development of policies for environmental protection, Aboriginal land rights, and appropriate safeguards.[1] Thus, as worldwide exploration and production activity surged in response to the post-1973 explosion in uranium prices, the industry in Australia stagnated.

The election of a Liberal government in 1976 and the publication of the Ranger Uranium Environmental Inquiry Report produced an atmosphere more conducive to the expansion of the uranium mining industry. Mary Kathleen, which had been placed on a "care and maintenance" basis following its closing in October 1963, was reopened in 1974, and exports of yellowcake began in late 1976. Progress on the "new" Northern Territory deposits, however, was further delayed by protracted negotiations with the Northern Lands Council (NLC), representing Aboriginal interests in the

Northern Territory, and stringent environmental requirements. By the time the Nabarlek and Ranger deposits commenced production (in 1980 and 1981, respectively), the boom in uranium prices had been and gone.

The Alligator Rivers region in the Northern Territory is estimated to contain about 81 percent of Australia's known reserves, i.e., approximately 330,000 tons U_3O_8.[2] Battey conjectures that "the potential of this province is of the order of five to ten times the known reserves."[3] The uncertainty surrounding the industry during the mid-1970s, however, discouraged more detailed quantification of reserves at existing deposits. In an attempt to remove this uncertainty and encourage the mining and export of uranium, in 1978 the federal government passed enabling legislation supporting the main elements of its uranium-development policy for the Northern Territory. This legislation was concerned chiefly with environmental protection and Aboriginal land rights and was largely based on proposals put forward by the Ranger Inquiry. Henceforth, government approval of mining operations would be forthcoming provided that adequate environmental protection measures were laid down by the mining companies, and that agreement was reached with the Northern Land Council concerning the preservation of Aborginal sacred sites. Uranium-mining projects in Queensland, South Australia, and Western Australia are not subject to the same degree of legislation, although ultimately the federal government controls the industry nationwide through its export controls.

There is no restriction on the amount of foreign ownership of uranium-exploration projects, but a minimum of 75 percent Australian equity *and* 75 percent Australian control is required of all projects entering the development stage. Prior to 1976, 100 percent Australian ownership was required. The locations of the major uranium deposits in Australia are shown in Figure 6.1; their current ownership and estimated reserves are summarized in Table 6.3.

A vast new copper/uranium/gold deposit is currently being evaluated at Olympic Dam on Roxby Downs in South Australia. While the project is still in the pilot stage, the joint venturers have estimated that uranium resources are in the vicinity of 1.3 million tons U_3O_8. In addition, there is an estimated 12 million tons of copper metal and commercial quantities of gold and silver. The viability of the project will depend on the demand for copper, uranium being the secondary product. While the deposit is of relatively low ore grade (averaging 0.06 percent U_3O_8), if uranium is mined as a co-product its extraction may prove relatively economical. Cost studies are currently being undertaken with a preliminary production estimate of about 165,000 tons of copper and 3,300 tons U_3O_8 a year, beginning around 1990.

If the uranium resources at Olympic Dam prove to be economically viable (i.e., if they can be classified as uranium reserves), then current WOCA uranium reserves will be boosted by 50 percent. As a further illustration of the size of the resource, consider annual WOCA uranium consumption requirements during the 1990s. For reactors currently operational, under construction, or on firm order, these requirements are estimated to average approximately 65,000 tons U_3O_8 a year over the decade, at which rate of consumption Olympic Dam could provide the equivalent of

FIGURE 6.1 Uranium Deposits and Prospects of Australia.
Source: OECD Nuclear Energy Agency and the International Atomic Energy Agency. *Uranium: Resources, Production and Demand,* OECD, Paris, December 1983.

TABLE 6.3 Ownership and Reserves of Major Australian Uranium Deposits

Deposit	Ownership		Estimated Reserves (tons U_3O_8)	Planned Production (tons U_3O_8)
Ben Lomond (Q)	Total (F)[2]	100%	7,000	400
Beverley (SA)	Western Nuclear (US)[3]	50%		
	Oilmin (Aus)	16.66%	17,000	—
	Petromin (Aus)	16.66%		
	Transoil (Aus)	16.66%		
Honeymoon (SA)	Mines Administration (Aus)[4]	34.23%	4,000	500
	Carpentaria Exploration (Aus)[5]	65.77%		
Jabiluka (NT)	Pancontinental Mining (Aus)	65%	230,000	5,000
	Getty Development (US)[6]	35%		
Koongarra (NT)	Denison Australia (Can)[7]	100%	15,000	2,200
Lake Way (WA)	Delhi Petroleum (Aus)[8]	53.5%	7,500	550
	Vam (Aus)	46.5%		
Mary Kathleen[1] (Q)	Conzinc Riotinto of Australia (Aus)[9]	51%	—	948*
	Australian federal government	41.64%		
	Australian public	7.36%		
Maureen (Q)	Central Coast Exploration (Aus)	51%	4,000	—
	Getty Mining (US)[6]	49%		
Nabarlek (NT)	Queensland Mines (Aus)[10]	100%	13,000	1,418*

Note: NT = Northern Territory, Q = Queensland, SA = South Australia, WA = West Australia, Aus = Australia, US = United States, Can = Canada, UK = United Kingdom, FRG = Federal Republic of Germany, F = France.

*Actual production in 1982.

[1] Mary Kathleen was shut down permanently in September 1982.

[2] A wholly owned subsidiary of Compagnie Française des Pétroles.

[3] A wholly owned subsidiary of Phelps Dodge.

[4] A wholly owned subsidiary of CSR.

[5] A wholly owned subsidiary of MIM Holdings.

[6] A wholly owned subsidiary of Getty Oil.

[7] A wholly owned subsidiary of Denison Mines.

[8] A wholly owned subsidiary of CSR.

[9] Rio Tinto Zinc (UK) owns 57.2 percent of CRA; the Australian public holds the remaining shares.

[10] A wholly owned subsidiary of Pioneer Concrete Services.

TABLE 6.3 (*Continued*)

Deposit	Ownership		Estimated Reserves (tons U_3O_8)	Planned Production (tons U_3O_8)
Olympic Dam (SA)	Western Mining (Aus)	51%	400,000[12]	3,300
	BP Australia (UK)	49%		
Ranger (NT)	Energy Resources Australia (Aus)[11]	100%	140,000	3,455*
Yeelirrie (WA)	Western Mining (Aus)	90%	50,000	2,750
	Urangesellschaft Australia (FRG)	10%		

[11] Ownership is Electrolytic-Zinc Company of Australia (Aus) 30.85 percent; Peko-Wallsend Operations (Aus) 30.85 percent; Australian public 13.3 percent; Japan Australia Uranium Resources Development (Japan) 10 percent; Rheinbraun Australia (FRG) 6.25 percent; UG Australia Developments (FRG) 4 percent; Interuranium Australia (FRG) 3.75 percent; Oskarshamnsverkets Kraftsgrupp Aktiebolag (Sweden) 1 percent.

[12] Not included in the OECD/IAEA RAR estimate for Australia.

about 20 years' supply. In terms of 1982 WOCA uranium consumption, this resource could provide at least 30 years of requirements at that level.

OECD/IAEA estimates made in 1979 envisaged a prosperous future for Australian uranium producers. Production was forecast to reach 15,500 tons U_3O_8 by 1985, and 26,000 tons (or 17 percent of WOCA production) by 1990. Assuming a price of $30/lb (which is slightly less than the average contract price received for Australian uranium in 1982), this would have represented annual export receipts (at 1982 prices) of $960 million in 1985 and $1,600 million in 1990. This would probably have ranked uranium in Australia's top three in terms of annual mineral export revenue earnings during the 1980s.

Just two years later this apparent bonanza had vanished! The corresponding OECD/IAEA forecasts made in 1981 estimated that Australian uranium production would rise to just 5,800 tons U_3O_8 in the early 1980s, peak at 7,800 tons in the mid-1980s, and then fall to 6,100 tons by the end of the decade. While this dramatic revision of OECD/IAEA production estimates was not peculiar to Australia, the decline was predicted to be greater there as the industry was still in its infancy. The 1983 OECD/IAEA estimates for Australia given in Table 6.1 project a further decline in production capability.

Australia has no commercially operating, or planned, nuclear-power reactors, and, as it is a nation well endowed with low-cost reserves of coal, this position is unlikely to change in the foreseeable future. Domestic consumption of uranium is therefore negligible and likely to remain so. A federal election in 1983 saw the Australian Labor Party win office with a uranium-mining policy that is unequivocally opposed to the expansion of the domestic industry, although it has undertaken to honor existing export contracts from the two operational deposits of Nabarlek and Ranger. Thus prospects for an expanding uranium-mining industry in Australia during the 1980s are extremely bleak. Because copper is its primary product, development of the Olympic Dam project has received government approval and uranium exports (beginning around 1990) from this deposit will be permitted.

Nabarlek has export contracts with two Japanese utilities (both of which provided finance to fund the project), the Commissariat à l'Energie Atomique (France), and a Finnish utility amounting to about 65 percent of current mill capacity (all ore has been mined and stockpiled). Ranger has more than 95 percent of its initial planned production to 1996 under contract with U.S., Japanese, South Korean, and Western European buyers (many of whom are equity holders). If justified by demand, production from the Ranger deposit could be doubled relatively quickly.

Production of uranium in Australia in 1982 totaled 5,821 tons U_3O_8 (1981 = 3,700 tons) with exports amounting to 6,860 tons (1981 = 1,790 tons) for a total export revenue of about A$490 million (1981 = A$120 million). This represented an average 1982 export price of A$35.68/lb U_3O_8 (or US$36.30/lb U_3O_8).

CANADA

Between 1933 and 1940 radium was recovered from the Bear Lake pitchblende deposit in the Northwest Territories, and this mine was reopened in 1942, at the request of the U.S. government, to provide uranium for the Manhattan project. In the same year, and for the same purpose, the Eldorado Gold Mining Company started a uranium-exploration program. In 1944 this company was acquired by the Canadian government, and a Crown company, Eldorado

Mining and Refining (subsequently known as Eldorado Nuclear, and currently as Eldorado Resources), was formed. A ban on private prospecting for radioactive materials was lifted in 1947, and various incentives were offered by the federal government in an effort to encourage exploration. By 1959, 23 mines with 19 treatment plants were in operation in five producing districts, with the majority of the uranium being produced in the Beaverlodge area of northern Saskatchewan, the Elliot Lake district in northern Ontario, and the Bancroft area of southeast Ontario.

The abrupt decline in uranium demand in the United States in the late 1950s saw uranium exploration virtually cease in Canada, while production fell sharply from a record 15,890 tons U_3O_8 in 1959 to 3,700 tons in 1968. By 1968 only four uranium companies were still in business, largely due to stretchout programs and a government stockpiling program which, between 1963 and 1970, purchased about 9,100 tons U_3O_8 at a total cost of C$101.4 million. The 1966 U.S. embargo on the enrichment of foreign uranium for use by domestic electricity-generating utilities was particularly severe on the Canadian uranium industry, which was heavily dependent upon the U.S. market.

When the market recovered in the mid-1970s, established Canadian producers were better placed to take advantage of the rapid surge in prices than their Australian counterparts. Between 1973 and 1980 Canadian production almost doubled (from 4,760 tons U_3O_8 to 9,290 tons), whereas Australian production was frozen pending the outcome of the Ranger uranium inquiry and negotiations with the Northern Land Council. Controversial issues surrounding the mining of uranium in both Australia and Canada (such as the question of native land-rights, nuclear proliferation, and environmental protection) caused lengthy delays in the development of the fledgling Australian uranium industry, whereas only new Canadian developments were affected, and, even then, these issues were resolved more expeditiously than in Australia.

In 1977 the Canadian government placed a temporary embargo on deliveries of uranium to the European Economic Community and Japan while it considered a formal policy relating to the long-term security of uranium for domestic use, in addition to pricing, reprocessing, and proliferation issues. It resolved that:

1. There must be a 30-year reserve requirement for existing, committed, or planned reactors in Canada before export permits were to be permitted.

2. Exports must be made at the world price or a floor price plus escalation, whichever is higher, and if possible must be upgraded to UF_6 prior to leaving Canada.
3. Appropriate safeguards regarding the use of Canadian uranium were to be agreed to by prospective customers.

The locations of Canada's uranium deposits are shown in Figure 6.2. In 1983, 69 percent of production came from the Elliot Lake deposits in Ontario, with the remaining 31 percent from

FIGURE 6.2 Uranium Deposits in North America.
Source: OECD Nuclear Energy Agency and the International Atomic Energy Agency. *Uranium: Resources, Production and Demand,* OECD, Paris, December 1983.

northern Saskatchewan. While the development of new deposits in northern Saskatchewan during the 1980s will undoubtedly raise its share of total production (to around 60 percent by 1990), the huge reserves at Elliot Lake mean that this area will maintain its importance as a major uranium producer for at least the next 30 years. Significant deposits are also known to exist in British Columbia, but exploration activity was curtailed in 1980 by a seven-year moratorium on uranium exploration and mining imposed by the provincial government. A moratorium on uranium exploration has also been imposed in Nova Scotia.

The development of the large, high-grade, deposits discovered in the mid-1970s at Cluff Lake and Key Lake in the Athabasca Basin region of northern Saskatchewan was the subject of major provincial government inquiries before development approval was forthcoming. The development of both projects was subjected to a long list of conditions regarding environmental protection, employment opportunities for local native labor, the investigation of native land claims, and the provision of financial aid to northern Saskatchewan to enable the local inhabitants to share the benefits of the proposed mines.

Many Canadian politicians believe that the Canadian economy is excessively dependent on exports of its natural resources and that foreign (mainly U.S.) ownership of these resources is too great. In the case of uranium, there are no restrictions on foreign investment in Canada's uranium industry at the exploration stage, but new production operations are required by federal government policy to have no more than one-third foreign ownership. In exceptional circumstances this could be raised to 50 percent. The current ownership and levels of production in 1982 of Canada's major uranium projects are summarized in Table 6.4.

Almost 50 percent of Canada's reserves (RAR) are contained in the Elliot Lake and Agnew Lake areas of Ontario. Ore grades in these areas, however, are low, generally averaging about 0.10 percent U_3O_8. Most of the remaining reserves are in northern Saskatchewan where ore grades are, on the average, considerably higher. The Key Lake deposit, where production officially began in September 1983, is expected to have an annual production rate of 4,000 tons U_3O_8 when it attains full capacity. This would make it second only to the Rossing mine in Namibia in terms of annual mine production of U_3O_8, pushing Australia's Ranger mine into third place. Key Lake ore will initially average 2.50 percent U_3O_8, while the average grade at the "D" deposit at neighboring Cluff Lake was 7 percent. Ore grades at this latter deposit (all ore from

TABLE 6.4 Ownership of Operational Canadian Uranium Projects

Location	Ownership		Annual Production Rate (tons U_3O_8) 1982
Agnew Lake[1] (O)	Kerr-Addison (C)	90%	90
	Uranerz (FRG)	10%	
Bancroft[2] (O)	Madawaski Mines (C)	100%	200
Beverlodge[3] (S)	Eldorado Nuclear (C)	100%	360
Cluff Lake (S)	Amok[4] (F)	80%	2,000
	SMDC[5] (C)	20%	
Elliot Lake (O)	Denison Mines (C)	100%	3,000
Elliot Lake (O)	Rio Algom[6] (C)	100%	3,400
Key Lake (S)	SMDC (C)	50%	
	Uranerz (FRG)	33.33%	4,000[7]
	Eldorado Resources (C)	16.67%	
Rabbit Lake (S)	Eldorado Resources (C)	100%	1,450

Note: O = Ontario, S = Saskatchewan, C = Canada, FRG = Federal Republic of Germany, F = France.

[1] Mining was suspended in 1979, but a surface salvage leaching operation continued into 1982. All operations ceased in early 1983.

[2] The mine was placed on standby in mid-1982 following the termination of a long-term contract with the Italian state-owned company, Agip, and was shut down early in 1983.

[3] After 30 years of production, the mine was closed during 1982.

[4] Ownership is Compagnie Française de Mokta 37%
Commissariat à l'Energie Atomic 30%
Péchiney-Ugine-Kulmann 25%
Cogema 8%

[5] Saskatchewan Mining Development Corporation.

[6] Rio Tinto Zinc (U.K.) owns 52.75% of Rio Algom.

[7] Planned annual rate at full capacity. Initial production began in September 1983.

which has now been completely mined and stockpiled) sometimes reached 45 percent, and special protective shielding was required for the mine operators.

Although the Elliot Lake deposits are of relatively low grade and high cost, Denison Mines and Rio Algom have received substantial orders from the Ontario Hydro Electric Company, which have ensured their existence for the next 30 years. Denison Mines has considerably expanded its mill capacity despite objections from Ontario Hydro that such action was not warranted by current market conditions. Denison's expansion, combined with the current and potential production capabilities of the Cluff and Key Lakes deposits, could push annual production to 19,000 tons U_3O_8

by the mid-1980s, although in view of current market conditions this is unlikely. In theory, annual production could reach 22,000 tons by 1990.

At year-end 1982, Canada had 4.7 GWe of installed nuclear capacity, all in Ontario. Early in 1983, two new nuclear-power plants became operational, one in New Brunswick, the other in Quebec, bringing total capacity to almost 6 GWe. By 1991 the 8.1 GWe of nuclear capacity now under construction in Ontario will raise this total to about 14 GWe. These figures represent a growth in annual uranium consumption from 1,350 tons U_3O_8 in 1983 (or about 13 percent of Canadian production in that year) to approximately 2,000 tons by 1992.

The bulk of Canada's current and long-term (up to 30 years for some contracts) domestic uranium requirements are under contract with the Elliot Lake producers, Denison Mines and Rio Algom. The former also has a large contract (20,000 tons U_3O_8 over 10 years beginning in 1984) with Tokyo Electric Power, and no further commitments are envisaged. The remainder of Rio Algom's production over the near future is also largely under contract, mainly with FRG, Japanese, South Korean, and U.S. utilities. The major French equity interests in the Cluff Lake project have ensured a substantial market for its output, with additional sales to the Federal Republic of Germany. Although the Key Lake project was developed at the time of a depressed market, it has nevertheless managed to achieve sufficient contracts to 1990 to achieve a "base loading."

Canada's past and future committed exports of uranium are more than those of any other nation. Since commercial contracting began in 1966, Canadian producers (by year-end 1983) had entered into arrangements to export about 147,000 tons U_3O_8, of which about 60,000 tons had already been exported. Thus forward-export commitments amount to about 87,000 tons U_3O_8, while about 104,000 tons has been committed for domestic use. Production in 1983 was about 9,400 tons, a 10 percent decrease over the 1982 figure of 10,500 tons. Canadian production should exceed 13,000 tons in 1984 with Key Lake operating throughout the year and the start-up of an additional uranium mill (Stanleigh) at the Rio Algom deposit. In 1984 not only should Saskatchewan replace Ontario as Canada's largest uranium-producing province, but Canada should replace the United States as the world's leading uranium producer.

The 1983 average weighted price for all export contracts made by Canadian producers for deliveries in 1983 was approximately

US$30.40/lb U_3O_8. Spot sales accounted for only about 1 percent of total exports.

FRANCE

Uranium prospecting in France began in 1946. Since then only the United States has spent more on domestic and foreign uranium exploration as the French have sought to achieve an independent nuclear arsenal and fuel for one of the world's major nuclear-power expansion programs.

The dominant company in all aspects of the nuclear fuel cycle in France is COGEMA, a 100 percent subsidiary of the French government's Commissariat a l'Energie Atomique (CEA). Four industrial groups are also involved in exploration and mining: Imetal—an association whose share capital is divided equally between Péchiney-Ugine-Kulmann (PUK) and Compagnie Française des Pétroles (CFP)—Minatome, Société Nationale Elf-Aquitaine (SNEA), and Rhone-Poulenc. The extent of their ownership in 1982 of France's major uranium (exploration, mining, and concentration) companies is shown in Table 6.5.

In general, individual uranium deposits in France are small and of relatively low grade. The domestic mining sector comprises a number of properties in Brittany, the Véndee, Massif Central, and Languedoc. The prospects for expansion of both resources and production, however, are limited, and consequently by 1990 domestic production will represent only about one-third of France's domestic uranium requirements. The comparable figure for 1982 was about 80 percent. With the prospect of a heavy reliance on imported uranium to meet their expansionary nuclear program, French companies have been extremely active in overseas ventures, especially in France's former territories of Gabon and Niger. In addition, four French companies (through Amok Ltd.) have the major shareholding in Canada's rich Cluff Lake deposit, while several French companies have shown interest in possible participation in the development of new Australian deposits.

Uranium production in France totaled about 3,770 tons U_3O_8 in 1983, this total being augmented by a similar quantity from Gabon and Niger. France reexports uranium through the corporation Uranex.

TABLE 6.5 Ownership of the Uranium-Mining Industry in France (1982)

Company	Group	Share of Capital (%)	Ore Concentration Plant
Compagnie Française de Mokta (C.F.M.)	IMETAL	SIMURA[4] 51 SMUC[5] 33.3	Le Cellier
Compagnie Generale des Matières Nuclèaires (COGEMA)	C.E.A.[1]	SIMO[6] 100 SMUC 33.3	Bessines, Ecarpiere St. Martin du Bosc
Compagnie Industrielle et Minière (C.I.M.)	Phone-Poulenc	——	——
Compagnie Minière Dong Trieu	Schneider S.A.	——	Mailhac-sur-Benaise
Total	C.F.P.[2] 50% P.U.K.[3] 50%	SCUMRA[7] 94	St. Pierre du Cantal
Société Nationale Elf-Aquitaine Productions (S.N.E.A.P)			

[1] Commissariat à l'Energie Atomique
[2] Compagnie Française des Pétroles
[3] Péchiney-Ugine-Kulmann
[4] Société Industrielle et Minière de l'Uranium
[5] Société des Mines d'Uranium du Centre
[6] Société Industrielle des Minerais de l'Ouest.
[7] Société Centrale de l'Uranium et des Minerais et Métaux Radioactifs

Source: OECD Nuclear Energy Agency and the International Atomic Energy Agency, *Uranium Resources, Production and Demand,* OECD, Paris, December 1983.

GABON

Uranium exploration in Gabon began in 1948, and the Mounana deposit (see Figure 6.3) was discovered in December 1956. The Compagnie des Mines d'Uranium de Francoville (COMUF) was established in 1958 to mine the Mounana deposit, and currently it is the only uranium producer in Gabon. Production in 1982 was about 1,200 tons U_3O_8. COMUF is jointly owned by the government of Gabon and a consortium of French companies. Shares are as follows:

Government of Gabon	24.75%
Cogema	18.81%
Total	13.00%
CFM	39.98%
COMUF employees	0.99%
Comp. de Générale D'Investissement	2.47%

(Details of ownership of the above French companies are given in Table 6.5. The government of Gabon has recently announced its intention to increase its holding in COMUF to over 25 percent.)

FIGURE 6.3 Uranium Deposits and Occurrences in Africa.
Source: OECD Nuclear Energy Agency and the International Atomic Energy Agency. *Uranium: Resources, Production and Demand,* OECD, Paris, December 1983.

In addition to the French (through Cogema), the Japanese Power Nuclear Fuel Corporation and the Korean Electric Power Corporation are actively involved in exploration in association with the government of Gabon.

A yellowcake plant started operations in 1978, prior to which all ore was shipped to France. France remains the dominant buyer, with small amounts being shipped to Italy and Japan. As in Niger, infrastructural problems cause the ore to be of relatively high cost.

The current status of Gabon's uranium deposits is as follows:

Deposit	Year of discovery	Ore content (%)	RAR (<$30/lb U_3O_8) (tons U_3O_8)	Status
Mounana	1956	0.48	n.a.	Mined out
Mikovloungou	1965	0.352	n.a.	High cost resource
Boyindzi	1967	0.407	3,600	Mining commencing
Oklo	1968	0.420	11,000	Mining in progress
Okelobondo	1974	0.436	9,600	
total			24,200	

Source: OECD Nuclear Energy Agency and the International Atomic Energy Agency. *Uranium: Resources, Production, and Demand,* OECD, Paris, December 1983.

NAMIBIA

Namibia is rich in mineral resources, and mining accounts for approximately 50 percent of gross domestic product and 70 percent of export earnings. Diamonds are the major extractive industry, with uranium ranked second. Cadmium, copper, lead, manganese, silver, tin, tungsten, and zinc are also important minerals in the Namibian economy.

Currently there is only one mine, Rossing, producing uranium in Namibia. The Rossing deposit was discovered in 1928, but its low ore grade made it an uneconomic proposition. Extensive prospecting activities that began in 1966, when the British company Rio Tinto Zinc (RTZ) acquired the exploration rights, culminated in the establishment of the world's largest open-cut uranium mine in 1975. While the ore grade is low (averaging around 0.04–0.05 percent U_3O_8), the massive scale of the project (a total of 300 million tons of ore) allows this open-cut mining venture to reap

TABLE 6.6 Ownership of the Rossing Deposit in Namibia

Company	Share (%)
Rio Tinto Zinc (UK)	46.50
Rio Algom (C)[1]	10.00
GENCOR (SA)	2.30
Industrial Development Corporation[2] (SA)	13.47
Total (F)[3]	10.00
Others (unknown)	17.73

Note: UK = United Kingdom, C = Canada, SA = South Africa, F = France.
[1] Rio Tinto Zinc owns 52.75% of Rio Algom.
[2] A state-owned corporation.
[3] A wholly owned subsidiary of Compagnie Française des Pétroles (F).

considerable economies of scale. Early problems with the abrasive nature of the ore and a fire in the process plant restricted production prior to 1979, but Rossing was due to reach design capacity of 5,000 tons U_3O_8 in 1983. Production in 1982 was 4,900 tons. Rossing's reserves are estimated to be about 135,000 tons U_3O_8.

Current ownership of the mine is given in Table 6.6.

South African-controlled uranium-mining ventures are generally veiled in secrecy. The "others" in Table 6.6 probably include a South African interest and was rumored at one time to have included an Iranian interest. Contract agreements are also secret, although its major customer in the past has been the United Kingdom through RTZ. Shipments have also been made to France, Japan (the latter through an agreement with RTZ), and Taiwan.

A second Namibian uranium deposit, at Langer Heinrich, is operating a pilot plant. This deposit, which is close to Rossing, is controlled by South Africa's General Mining Union Corporation (GENCOR), but details have not been disclosed.

Apart from machinery problems, the operation at Rossing has been hampered by labor unrest, generally with regard to different wage scales for different races. While Rossing is supposed to comply with South African apartheid policy, the management has in general ignored it, seeking a nonracial operation. A far greater problem, however, is political unrest and pressure on major uranium-consuming nations not to purchase uranium from (or invest in) a country that is being controlled against its will by South Africa. The government of the United Kingdom has decided not to contract for any more uranium from Namibia for delivery after 1984. Unless a realistic settlement to the question of nationhood for Namibia can be reached in the near future, exploration and investment in the uranium industry (and probably most other industries) is likely to be severely discouraged.

NIGER

Up to 1974, Niger's main export money-earner was peanuts, but a severe drought in the mid-1970s destroyed this source of income entirely. Until the mid-1970s Niger had a chronic balance-of-trade deficit, but the expansion of its uranium trade during the price boom of the mid-1970s turned this into a small surplus. In 1983, uranium exports accounted for about 66 percent (by value) of total exports, or about one-fifth of the national budget. In turn, imports are greatly influenced by the requirements for capital equipment, sulfur, and fuel for the uranium industry.

In the mid-1950s the French Atomic Energy Commission made surveys of the Air district in north-central Niger, although it was not until 1965 that the first economically viable deposits were found in the Arlit region. Production did not begin until 1971. Exploration and development work have accelerated ever since, until today the area west of the Air mountains—stretching for nearly 200 km—comprises one of the world's most prolific uranium locations. The high cost of exploration, however, has prevented detailed enumeration of Niger's uranium resources, with Koutoubi and Koch[4] giving a range of from 130,000 to nearly 650,000 tons U_3O_8.

Niger has two uranium facilities, Arlit and Akouta, which produced a combined annual total of about 4,600 tons U_3O_8 in 1982. All exploration, production, and marketing of mineral resources in Niger is the responsibility of Onarem (Office National des Ressources Minières), a government institution organized like a private commercial company. Onarem can participate in all companies or groups engaged in exploration or mining activities in Niger and currently controls about one-third of the country's uranium production. France, through Cogema and other French shareholders, has a substantial share of the equity in the two projects that are currently operational. Details of these two projects and the Arni deposit, development of which has been delayed due to depressed market conditions, are given in Table 6.7.

With the exception of Onarem, Niger's uranium production has in the past been committed to the equity partners on a pro-rata

TABLE 6.7 Niger's Uranium Deposits[1]

Company (deposit)	Ownership	
Somair[2] (Arlit)	Onarem (N)	33%
	Cogema (F)[5]	26.96%
	Total (F)[6]	26.96%
	Urangesellschaft (FRG)	6.54%
	Agip (I)[7]	6.54%
Cominak[3] (Akouta)	Cogema (F)	34%
	Onarem (N)	31%
	OURD (J)[8]	25%
	Enusa (S)[9]	10%
SMTT[4] (Arni)	Onarem (N)	33.33%
	Cogema (F)	33.33%
	KFTC (K)[10]	33.33%

Note: N = Niger, F = France, FRG = Federal Republic of Germany, I = Italy, J = Japan, S = Spain, K = Kuwait.

[1] Plans for the early development of a fourth mine, Imouraren, have been shelved pending an improvement in the uranium market. Feasibility studies are continuing at a number of other promising uranium finds.

[2] Société des Mines de l'Air.

[3] Compagnie Minière d'Akouta.

[4] Société Minière de Tassa N'Taghalgué.

[5] A wholly owned subsidiary of France's Commissariat à l'Energie Atomique (CEA).

[6] A wholly owned subsidiary of Compagnie Française des Pétroles.

[7] An agency of the Italian government.

[8] Overseas Uranium Resources Development Company.

[9] Empresa Nacional del Uranio.

[10] Kuwait Foreign Trade and Contracting Company.

basis. Onarem is not required to take all of its share, and it appears that French interests have in the past absorbed any residual. More recently, however, Onarem has become more independent and has made small sales to a number of different buyers. A substantial sale to Libya, totaling approximately 1,500 tons U_3O_8, in early 1981 caused great concern to some Western governments, inasmuch as Libya has no legitimate use for unenriched uranium. It is rumored that sales have also been made to Iraq and Pakistan, both of which are thought to be actively involved in constructing nuclear weapons. Niger has no safeguards policy, and such sales would attract a premium in excess of current market prices. In addition, Niger cannot afford to offend its powerful northern neighbor, Libya.

Niger is a high-cost (i.e., over \$30/lb U_3O_8) producer of uranium. The mines are situated in a desolate and remote area and, while the infrastructure has been substantially upgraded by the building of an all-weather, sealed "Uranium Road" linking the inhabited south of the country to the mining areas 650 km to the north, the costs associated with transporting uranium to its overseas markets and importing fuel and mining equipment are substantial.

Prospects for Niger's uranium industry appear reasonably good provided that the French maintain their high-cost (relative to, say, Australian or Canadian yellowcake) purchases. Uranium sales, and the foreign currency they provide, are vital to the economic and political stability of Niger, and the French can undoubtedly justify the cost of their uranium in terms of foreign aid to this impoverished nation.

SOUTH AFRICA

Active exploration for uranium in South Africa began in the late 1940s and culminated in the large-scale production of uranium oxide as a by-product of the gold mining industry in 1952. By 1960 production had reached 6,400 tons U_3O_8 (approximately 16 percent of world production in that year), with the U.K. and U.S. weaponry programs being the major customers. Thereafter, production dropped rapidly, and it was not until the late 1970s that it passed the 1960 level.

Gold is South Africa's primary commodity, accounting for almost half of all exports in recent years. Other minerals of importance are coal, diamonds, iron ore, copper, and manganese. Uranium is of relatively minor importance, accounting for only about 2 percent of exports.

Virtually all of South Africa's uranium is produced as a by-product of gold mining in the Witwatersrand basin (see Figure 6.3). As a consequence, its recovery costs are very low. Furthermore, South Africa's gold producers attempt to maximize the life of their mines by mining the marginal grades of ore. If gold prices rise, lower-grade ore that has now become profitable will be mined. Thus, although ore production may increase as a result of the price increase, the amount of gold recovered may actually decline. Since uranium and gold appear in the ore in a fairly constant ratio, it follows that as the price of gold rises, uranium production may also fall.

Slimes (tailings) dams resulting from the operations of gold and gold/uranium mines contain low concentrations of gold, uranium, and pyrite. The high price of these three minerals in the late-1970s encouraged their extraction from the slimes dams. In 1982 three such projects were operational.

Uranium is also produced as a (minor) by-product from the Palabora open-cut copper mine.

Two mines, Beisa and Afrikander Lease, were scheduled to come on-stream in 1982 as the first primarily uranium mines (with gold as a by-product) in South Africa. The development of Beisa was completed, but the Afrikander Lease awaits recovery of the uranium market. The Beisa mine, however, is scheduled to be shut down by the end of 1984, due to the depressed state of the uranium market.

The major mines and mining houses engaged in uranium recovery are shown in Table 6.8 together with their levels of uranium production in 1982. Overall, the average recovery grade is very low, approximately 0.01 percent U_3O_8, reflecting the by-product nature of South African uranium mining.

With the exception of Palabora's uranium production, all South African uranium is marketed by NUFCOR, a private company managed by the country's major uranium producers. South Africa's Atomic Energy Act forbids the release of details concerning uranium contracts, but in the past Japan, the Federal Republic of Germany, the United States, France, Taiwan, Belgium, and Spain have all been major customers. In an unusual break with the normal practice of secrecy, however, Japanese customers were identified as contracting for about 80 percent of the planned production from the Beisa mine, beginning in 1983. This action may have been intended to assure potential customers of reliable supplies in the face of mounting world pressure on the South African government over its apartheid policies and its apparent desire to retain control over Namibia. South African producers rely on spot market sales for a substantial amount of their production.

South Africa maintains a policy of racial segregation and separate "development." The majority of black South Africans are forced to live in their tribal homelands and are regarded as migrant workers in the mines. Most skilled and semi-skilled jobs are not open to nonwhites. As a consequence, in times of high demand for minerals a chronic shortage of skilled labor arises. In recent years, black workers have been permitted to form labor unions and their wages have been rising fairly rapidly as a result of both unionization and external pressure on South African companies that are

TABLE 6.8 South African Uranium Production (1982)

Mining Group[1]	Major Mines	Production (tons U_3O_8)
(A) *Uranium Deposits*		
Anglo American Corporation	Afrikander Lease[2]	—
GENCOR	Beisa (St. Helena)	280
	total	280
(B) *Gold-Mining By-product*		
Anglo American Corporation	Vaals Reef	1,898
	Western Deep Levels	202
Anglovaal	Hartebeestfontein	470
Barlow Rand	Blyvooruitzicht	278
	Harmony	652
Gold Fields of S.A.	Driefontein Cons.	250
GENCOR	Buffelsfontein	640
	West Rand Cons.[3]	—
Johannesburg Consolidated Investment	Randfontein	510
	Western Areas	190
	total	5,090
(C) *Tailings Mining Co- and By-product*		
Anglo American Corporation	East Rand Gold & Uranium	378
	Joint Metallurgical Scheme	952
GENCOR	CHEMWES	670
	total	2,000
(D) *Copper-Mining By-product*		
Palabora[4]	Palabora	190
	total	190
	Overall total	7,560

[1] Companies administering the projects. The ownership of individual mines is generally spread across a large number of companies. Cross-ownings are commonplace.

[2] Mining is suspended pending recovery of the uranium market.

[3] Uranium production ceased at the end of 1981 because of depressed market conditions.

[4] Rio Tinto Zinc (U.K.) owns 38.9 percent of Palabora, with the Newmont Mining Corporation (U.S.) being the other major shareholder (28.6 percent).

owned by overseas interests. Nevertheless, the gulf in earnings between blacks and whites remains considerable.

South Africa has been one of the world's major uranium suppliers for more than 30 years. Now that uranium is in relatively abundant supply, however, South Africa's continuing policy of apartheid and intransigence over Namibian independence may drive potential customers for uranium toward politically less controversial producers.

UNITED STATES

Radioactive metals were discovered in the western United States in the late 1880s, and subsequently the discovery of radium (in 1898) and its development for medical purposes and luminous paints generated a small mining boom in Colorado between 1912 and 1918. Thereafter the market was dominated by low-cost production from the Belgian Congo. Prior to the Second World War, uranium ore was also produced as a co-product of vanadium mining in the Colorado Plateau region, but less than 100 tons of yellowcake (primarily for use as paint or a glass coloring agent) was recovered.

The Manhattan project created the first real demand for uranium, most of which was met by supplies from the Belgian Congo and Canada. Following the establishment of the Atomic Energy Commission (AEC) in 1946, however, a nationwide search for uranium resources was launched, encouraged by generous discovery and development bonuses, together with guaranteed prices for uranium ore. Exploration activities were centered primarily in the Colorado Plateau states: Arizona, Colorado, New Mexico, Wyoming, and Utah. The AEC also encouraged exploration for uranium and the development of existing uranium deposits in many overseas countries.

The AEC's domestic program was extremely successful. Uranium production rose from less than 100 tons U_3O_8 in 1948 to more than 8,000 tons U_3O_8 in 1956. Although the AEC was the sole purchasing body for uranium ore, it encouraged private companies to enter the uranium-milling industry with extremely generous investment allowances. By the end of 1956 there were 12 privately owned uranium mills in operation, and this number had risen to 26 (processing 30,000 tons of ore a day) by 1962. No new mills were constructed over the ensuing 15 years, however, as the uranium industry entered the void between the decline in demand from the

U.S. nuclear-weapons program and the growth of the nuclear-power industry.

During the 1950s and early 1960s, Colorado and Utah were the leading states in uranium production, but their fortunes deteriorated in the mid-1960s and never recovered. Their places were taken by New Mexico and Wyoming, which have dominated production for the past 20 years. In 1981 New Mexico and Wyoming accounted for 55 percent of U.S. uranium production. When combined with Texas, these three states accounted for more than 70 percent of the total. The remainder is produced in Arizona, Colorado, Florida, Louisiana, South Dakota, Utah, and Washington. The bulk of low-cost ($<$50$/lb U_3O_8) reserves are also located in New Mexico and Wyoming, with shares of 43 percent and 33 percent, respectively, at the beginning of 1983. Areas of uranium ore deposits in the United States were shown in Figure 6.2. Production of uranium in Florida and Louisiana is as a by-product from phosphoric acid mills, and consequently these states are omitted from the map.

The degree of involvement of major oil companies at the mining and milling stages of production is very noticeable; they are owners or partners in about 50 percent of both activities. Gulf Oil, Continental, Getty, Exxon, Chevron, Conoco, and Phillips Petroleum all have substantial industry interests. The largest company in the uranium industry, however, is the mining company Kerr-McGee, which is engaged in the exploration, production, and marketing of oil, gas, and coal in addition to its widespread holdings in the uranium industry. In addition to owning 10 uranium mines (not all operational), Kerr-McGee also controls a UF_6 conversion facility. In 1982 its output of 2,105 tons U_3O_8 represented about 16 percent of U.S. production. United Nuclear and the French company, Cogema, also have substantial interests in the U.S. uranium industry.

Production figures for individual U.S. deposits are confidential, but an indication of the scale of involvement of companies in uranium mining can be obtained by noting their nominal daily milling capacity (Table 6.9). These figures are far from perfect as indicators of actual company production of yellowcake as they neglect the grade of ore processed, and hence actual output of U_3O_8 is unknown. In addition, no allowance is made for milling that is done on a toll basis for (generally small) mines without their own milling facility. Nevertheless, the previously mentioned involvement of oil companies in the U.S. uranium industry is immediately apparent from this table.

TABLE 6.9 U.S. Uranium-processing Plants (operating as of January 1, 1982)

Conventional Mills Company	Location	Nominal Capacity (tons ore/day)
Anaconda Minerals[1]	Grants, N. M.	6,000*
Atlas Minerals[2]	Moab, Utah	1,400
Bear Creek Uranium[3]	Powder River Basin, Wyo.	2,000
Chevron Resources[4]	Hobson, Tex.	2,500
Conoco-Pioneer Nuclear[5]	Falls City, Tex.	3,400*
Cotter[6]	Canon City, Colo.	1,200
Dawn Mining[7]	Ford, Wash.	450*
Energy Fuels Nuclear	Blanding, Utah	2,000**
Exxon Minerals[8]	Powder River Basin, Wyo.	3,200
Homestake Mining	Grants, N. M.	3,400
Kerr-McGee Nuclear[9]	Grants N. M.	7,000
Minerals Exploration[10]	Red Desert, Wyo.	3,000**
Pathfinder Mines[11]	Gas Hills, Wyo.	2,500
Pathfinder Mines	Shirley Basin, Wyo.	1,800
Petrotomics	Shirley Basin, Wyo.	1,500
Rio Algom[12]	La Sal, Utah	750
Union Carbide	Uravan, Colo.	1,300*
Union Carbide	Natrona County, Wyo.	1,400
United Nuclear[13]	Church Rock, N. M.	3,000*
Western Nuclear[14]	Wellpinit, Wash.	2,000*
total nominal capacity		49,800

Conventional Mills: total nominal capacity	19,000–21,000 tons U_3O_8/year
Solution Mining operations (largely in Texas and Wyoming)	1,700– 2,100 tons U_3O_8/year
Phosphoric Acid by-product (Florida and Louisiana)	800– 1,200 tons U_3O_8/year
Heap Leachings, Dumps, Tailings (Arizona, Colorado, Texas, Utah)	200–400 tons U_3O_8/year
	21,700–24,700 tons U_3O_8/year

[1] A wholly owned subsidiary of Atlantic Richfield Company (Arco).
[2] A division of the Atlas Corporation.
[3] Jointly owned by the Rocky Mountain Energy Company and Southern California Edison.
[4] A wholly owned subsidiary of Standard Oil Corporation of California.
[5] Jointly operated by Conco (a wholly owned subsidiary of Du Pont) and Pioneer Nuclear.
[6] A wholly owned subsidiary of Commonwealth Edison.
[7] Newmont Mining owns 51 percent of Dawn Mining.
[8] A division of Exxon Corporation.
[9] A division of Kerr-McGee Corporation.
[10] A wholly owned subsidiary of Union Oil Corporation of California.
[11] Jointly owned by Cogema (80 percent) and General Electric (20 percent).
[12] Rio Tino Zinc (UK) owns 52.75 percent of Rio Algom.
[13] A wholly owned subsidiary of Homestake Mining.
[14] A wholly owned subsidiary of Phelps Dodge.
*Plants not operating at year-end 1982.
**Plants shut down during 1983.

Source: Statistical Data of the Uranium Industry, U.S. Department of Energy, Grand Junction Colo., 1982.

As of January 1, 1981, a total of 22 conventional uranium mills were operating in the United States, with a total nominal daily capacity of 51,050 tons of ore. The following year, 20 mills were operating, with a nominal capacity of 49,800 tons ore/day (Table 6.9). By year-end 1982, however, the number of operating mills had fallen to 14, with a capacity of 33,700 tons ore/day, and the following year further closings forced capacity down to 27,450 tons ore/day. Thus over a period of just three years, U.S. uranium milling capacity fell by almost 50 percent, reflecting a rationalization of the industry in response to a weak uranium market and the availability of imports from low-cost producers in Australia, Canada, and to a limited extent South Africa.

In addition to the conventional mines recovering yellowcake from ore, in 1981 there were 11 solution-mining operations, seven plants recovering uranium as a by-product of phosphoric-acid production, and four plants using heap leaching to extract U_3O_8 from dumps or tailings. Combined, these 22 processing plants accounted for approximately 19 percent (i.e., 3,580 tons U_3O_8) of total U.S. production of uranium concentrate in that year.

By year-end 1981, cumulative (from 1947) U.S. uranium production had reached 365,963 tons U_3O_8, which was produced from more than 200 million tons of ore yielding an average grade of about 0.18 percent U_3O_8. As might be expected, the average grade of processed ore has fallen considerably over this period, from a high of 0.32 percent in 1952 to 0.20 percent in 1970, 0.16 percent in 1975, and 0.11 percent in 1979. Since 1979, however, cost pressures have forced many marginal (generally low-ore-grade) producers out of business, and consequently there was a small increase in ore grade to 0.12 percent in 1982. This trend can be expected to continue as U.S. producers attempt to compete with imports from the low-cost, high-grade, deposits in the Northern Territory (Australia) and Saskatchewan.

In 1980, U.S. uranium production reached an all-time peak of 21,840 tons U_3O_8 in (a lagged) response to the boom conditions of the late 1970s. The corresponding figures for 1981, 1982, and 1983, however, were 19,240 tons, 13,430 tons, and 10,600 tons, respectively, reflecting the rapid demise of many high-cost producers as the NUEXCO exchange value fell to its lowest level ever (in real terms) in early 1983. Given the lagged response to price changes that is inherent in the uranium mining industry, U.S. production is likely to continue its fall during the mid-1980s, although at a much reduced rate.

As an indication of the effects of the plunge in uranium prices on the mining sector since the late 1970s, Table 6.10 shows the dramatic fall that has occurred in U.S. domestic production potential for the years to 1992 as estimated (from survey data) by the DOE. In 1980 the survey indicated that production could reach 25,400 tons U_3O_8 by 1983, whereas the actual figure was 10,600 tons. The corresponding DOE figure for 1985 was 23,200 tons which, by its 1982 survey, had been revised to 14,300 tons. This amounts to a drop of almost 40 percent in just two years. These figures simply confirm the NUEXCO estimates in Table 6.2, as both foresee a rapid demise for many U.S. uranium-mining and -milling operations during the first half of the 1980s.

While U.S. uranium production is expected to decrease substantially during the 1980s, U.S. domestic consumption requirements are expected to increase considerably from about 10,500 tons U_3O_8 in 1982 to about 19,000 tons by 1990. Thus during the 1980s the United States will come to be more reliant on imported uranium to supplement its domestic supplies than at any time since the 1950s.

U.S. exports of (domestic source) uranium have never been large, amounting to a cumulative total of only 21,600 tons U_3O_8 between 1966 and 1982. Imports have also been of minor importance, amounting by year-end 1982 to a cumulative total of 14,500 tons since the import embargo was partly relaxed in 1977. Over the

TABLE 6.10 Total U.S. Domestic Production Potential (tons U_3O_8)

	Date of Survey			
Year	1980	1981	1982	1983
1983	25,400	21,600	18,000	14,550
1984	24,300	20,100	14,900	14,250
1985	23,200	19,200	14,300	15,050
1986		17,600	14,300	15,600
1987		15,700	13,700	14,600
1988		15,200	13,400	13,400
1989		13,400	13,400	12,950
1990		12,400	12,600	12,650
1991			11,900	11,550
1992				11,400

Source: Survey of United States Uranium Marketing Activity, U.S. Department of Energy, 1980–83.

next decade, U.S. exports will diminish to a negligible quantity while imports should rise at a relatively steady rate.

CONCLUDING REMARKS

U.S. production during the 1980s will be supplemented by a large inventory of uranium held as both U_3O_8 and enriched UF_6. At year-end 1982, total private and government-owned U.S. uranium inventory amounted to approximately nine years of forward consumption, which is excessive by any yardstick. While a certain amount of enriched UF_6 inventory is required by the DOE as a working inventory to assure enrichment-service customers that their deliveries can be met on contracted schedule, the high inventory of U_3O_8 has been reflected by depressed prices and consequently reduced levels of production.

Overly optimistic projections of electricity requirements made during the 1970s, combined with delays in licensing and constructing nuclear-power installations and an industrial recession, have produced a situation where uranium production has exceeded consumption by a substantial amount during the early 1980s, with the excess contributing to a largely unintended increase in utility inventories. Since these conditions have prevailed for a number of years, the uranium market is currently characterized by an excessive level of inventories, surplus production capacity, and, consequently, depressed prices.

Market forces, however, are now forcing an adjustment in production levels and a geographical redistribution of producing areas, with the United States in particular cutting back rapidly in the face of competition from Canada and Australia. When demand and supply eventually balance toward the end of the 1980s, new production facilities may be required. Until then, however, only the most cost-efficient operations are likely to survive the period of adjustment.

NOTES

1. Ranger Uranium Environmental Inquiry (1976).
2. This excludes speculative estimates of the size of the uranium reserves at Olympic Dam.
3. Battey (1978).
4. Koutoubi and Koch (1980).

7

The Demand for Uranium

INTRODUCTION

The demand for uranium is dependent upon the world's needs for primary energy, the proportion of that energy required as electricity, and the fraction of that electricity generated by nuclear power. Thus the demand for uranium is a derived demand and, ultimately, depends upon the relative costs and benefits (measured in terms of both economic and political variables) of the different modes of electricity generation. Projections of future levels of installed nuclear capacity (and hence uranium requirements), however, must necessarily incorporate changes in technology (both in the nuclear and nonnuclear power industries), fuel costs, long-run energy demand, and government policies. All four of these factors have changed frequently over the relatively short life-span of the nuclear-power industry, and forecasts of future levels of nuclear capacity have been subject to frequent, and often substantial, revisions. Forecasts of uranium requirements have been revised in a parallel manner.

Before discussing the numerous factors which, combined, determine the level of uranium demand, it is important to distinguish between demand and consumption at various stages of the nuclear fuel cycle.

Uranium demand is defined as uranium consumption in any given period plus the change (intended or unintended) in the level of uranium inventories. Consumption is defined as the amount of uranium entering the conversion–enrichment–fuel fabrication pipeline. Thus uranium is deemed to have been "consumed" before it reaches the reactor. It follows that inventories as defined here incorporate only producer and buyer inventories of U_3O_8. Inventories of UF_6 and enriched UF_6 are contained in consumption. These definitions of demand and consumption are reasonably standard throughout the nuclear-power industry, but it will be seen in Chapter 9 that they create a number of problems when it comes to the construction and estimation of an economic model of the U.S. uranium industry.

Reactor demand (or requirements) is defined as the actual amount of uranium used to fuel the nuclear reactor. If construction delays prevent the reactor from becoming operational according to schedule, or if the reactor must be shut down for an extended period, then clearly reactor demand is zero. Uranium will still be required, however, since conversion and enrichment contract schedules must still be met. Given the size and characteristics of nuclear-generating capacity that is expected to be operational over a given time horizon, it is possible to calculate "apparent future consumption," defined as the amount of uranium required to meet reactor demand over this given period.

The market for uranium can be considered a form of futures market, in the sense that it is possible (and in the past has been essential, to meet enrichment contract schedules) for utilities to enter contracts for the supply of uranium many years before its scheduled date of delivery. Given a life-span for nuclear reactors of approximately 30 years, it is possible for a utility to forecast fairly accurately future lifetime reactor consumption requirements (i.e., apparent future consumption) once the reactor has become operational. The amount of this "apparent future consumption" that is not already under contract or in the form of utility-held inventories is known as the reactor's "unfilled requirements." Unfilled requirements in aggregate (i.e., over all operational reactors in a given country) comprise the country's active market demand for uncommitted production.

Ultimately, the demand for uranium is determined by the amount of electricity generated now, and in the future, by nuclear-power reactors. Even if the nuclear-power industry were to experience a complete dry-up of reactor orders worldwide today, existing reactors and those under construction would ensure that its contribution to the total amount of electricity generated worldwide would increase from about 10 percent in 1982 to about 18 percent in 1990. For the developed OECD nations, these figures are 15 and 24 percent, respectively. In addition, since approximately 60 percent of the world's reactors that are currently in operation are less than 10 years old, the role of nuclear power as a major source of electricity is assured until well past the end of this century.

CHARACTERISTICS OF THE DEMAND FOR URANIUM

Uranium has only one major use, as an input in the fuel cycle for nuclear-power reactors. Other applications of uranium are so minor (and likely to remain so) that they can be ignored. Uranium is not unique in this respect; however, other similarly placed minerals, such as titanium, of which the major use is in the aerospace industry, generally have fairly close substitutes (aluminum and steel can be substituted for titanium in many applications). In addition, uranium concentrate is normally sold directly from the mine to the utility, which tolls the conversion, enrichment, and fabrication of the raw material. Thus changes in utility requirements are quickly transmitted to the mines (by way of contrast, consider the many and varied uses of aluminum and the number of intermediate stages between the bauxite mines and the final products).

The price of uranium is a very small component in the cost of nuclear-generated power. Eklund[1] has calculated that a doubling in the cost of the raw material would increase the cost of electricity generation by about 10 percent for nuclear power, and by 65 percent for fossil-fired power plants. Within the nuclear fuel cycle, the cost of processing the uranium accounts for approximately 88 percent of the cost of the final fuel assemblies, whereas comparable figures for coal and oil are 42 percent and 33 percent, respectively.

Uranium is relatively easy to store and transport. The annual fuel requirements of a 1,000-MWe nuclear-power plant would require less than 50 square meters of storage space, whereas the fuel storage space for comparable oil- and coal-fired plants would be 25 hectares and 40 hectares, respectively. Thus future requirements of uranium can to some extent be assured by carrying a high level of inventories, although this would involve a considerable opportunity cost during times of high storage costs (i.e., high interest rates). Inventories held by the world's electricity utilities, reactor manufacturers, and fuel fabricators at the beginning of 1983 have been estimated by NUEXCO to be 205,000 tons U_3O_8

equivalent,[2] although this figure is subject to a fairly substantial degree of uncertainty. To allow this figure to be put into context, it should be noted that it is approximately four times WOCA uranium production in 1982, and approximately five times WOCA uranium consumption in the same year.

GROWTH OF NUCLEAR POWER

During the late 1960s and early 1970s, nuclear power was heralded as the technology for providing industrialized nations with virtually unlimited supplies of low-cost electricity. Over this period, Japan, the United States and many Western European nations embarked on plans for a rapid expansion of their nuclear-generating capacity in response to projections of a continuing high rate of growth of their domestic economies and, consequently, in the demand for electricity. For example, prior to 1970, 90 nuclear plants, representing a capacity of 70.8 GWe, had been ordered in the United States. In the period 1970–74, an additional 110 plants with a planned capacity of 122.3 GWe were ordered.

The majority of plants ordered around 1970 were scheduled to enter service during the late 1970s and early 1980s with the result that installed nuclear capacity in WOCA countries was projected to rise from a level of 18 GWe in 1970 to 300 GWe by 1980 and 610 GWe by 1985. The corresponding projection (made in 1972) for 1990 was 1,068 GWe, thus implying that the rapid rate of plant ordering was expected to continue throughout the decade of the seventies and into the early eighties.

While the immediate impact of the OPEC oil embargo in 1973 was to strengthen the case for nuclear power for generating electricity in the major oil-importing nations of the developed world, the subsequent fall in industrial activity combined with energy conservation measures reduced the anticipated rate of growth in demand for electricity, and orders for new power plants (fossil as well as nuclear) were drastically reduced in most developed nations. The major threat to the long-term viability of the nuclear-power industry, however, came in the second half of the 1970s, when the industry became embroiled in numerous safety and environmental controversies, the major consequence of which was an increase in the amount of regulation governing the construction and operation of nuclear-power plants. This was particularly noticeable in the United States (especially in the wake of the accident at Three Mile Island on March 28, 1979), where the lead time between the awarding of a plant contract and commercial operation grew from about six years in the late 1960s to about 12 years by the end of the following decade. Such lengthy project times lead to cost increases and are generally longer than the traditional planning horizons in the utility industry. These issues, combined with high interest rates and an industrial recession, rendered nuclear power a politically and economically uncertain option for utilities, and forced the curtailment of expansion plans for nuclear power in many nations.

Despite the problems that have plagued the nuclear-power industry since the mid-1970s, installed nuclear capacity worldwide is projected to increase by 150 percent over the period 1982–90 (Table 7.1). Among OECD nations, just three countries (France, Japan, and the United States) currently account for 72 percent of the total level of installed nuclear capacity, and this percentage is expected to remain fairly constant through 1990. If Canada, the Federal Republic of Germany, Spain, Sweden, and the United Kingdom are added to this list, then these eight countries will account for almost 94 percent of the OECD total over the same period.

The highest rate of growth is projected to occur in the newly industrialized nations of Asia (South Korea and Taiwan) and South America (Argentina and Brazil), although in absolute terms the increase is relatively small. India is also planning a substantial increase in its nuclear capacity.

A threefold increase is projected for the centrally planned economies countries over the period 1982–90, with the USSR accounting for about three-quarters of this increase. The low–high range for this category in 1990, however, is extremely wide, reflecting the high degree of uncertainty inherent in these projections.

With France, Japan, the United States and the USSR accounting for about three-quarters of the projected growth in nuclear power during the decade of the 1980s, modification of existing expansion plans in any of these countries will have a marked effect on the totals given in Table 7.1. Previous projections of the rate of growth of installed nuclear capacity (and hence uranium demand) have been characterized by their frequent, and often substantial, downward revisions as it became apparent that earlier projections of electricity demand were too high. A series of projections of the growth of nuclear power published by the OECD and the IAEA since 1969 are given in Table 7.2 and are shown

TABLE 7.1 Nuclear-Power Growth Estimates (year-end, net GWe)

Country	1982[1]	1985	1990	2000
OECD Nations				
Austria	—	0.7	0.7	
Belgium	3.5	5.5	5.5	
Canada	7.0	10.3	14.5–16.9	
Finland	2.2	2.2	2.2	
France	23.8	39.0	59–67	
Germany (FRG)	9.9	17.0–18.3	25.0–29.3	
Greece	—	—	0.6	
Italy	1.3	1.4	7.4	
Japan	17.3	28.0–30.0	51–53	
Netherlands	0.5	0.5	0.5	
Spain	2.0	7.7	13.7–15.7	
Sweden	7.3	7.3	9.4	
Switzerland	1.9	2.9	4.8	
Turkey	—	—	0.6	
United Kingdom	6.1	9.4	12.7–13.5	
United States	63.9	86.0–109.0	123–139	
total	146.7	217.8–244.1	330.5–366.0	497–683
Developing Nations				
Argentina	0.4	0.97	1.67	
Brazil	0.626	0.63	5.61	
Egypt	—	—	0.6	
India	0.804	1.6–1.8	2.3–2.7	
Indonesia	—	—	0.6	
Korea (South)	1.185	3.59	9.89	

Mexico	—	1.3	2.3–3.1	
Pakistan	0.125	0.125	0.125	
Philippines	—	—	0.62	
Taiwan	3.2	4.02	4.9–6.7	
total	6.34	12.235–12.435	28.615–31.615	88–121
Centrally-Planned Economies[2]				
Bulgaria	1.8	1.62	1.62–2.57	
Cuba	—	—	0.41	
Czechoslovakia	0.88	3.20	5.72–6.68	
Germany (GDR)	1.8	3.33	4.97–5.92	
Hungary	0.44	1.61	1.61–2.56	
Poland	—	—	0.41–0.82	
Romania	—	—	1.30–1.92	
USSR	17.4	24–25	45–63	
Yugoslavia	—	0.63	0.63	
total	22.32	34.39–35.39	61.67–84.51	166–291

Other Nations

(1) South Africa is predicted to have 1.843 GWe of operational nuclear capacity by 1985.

(2) While the People's Republic of China is planning to establish a nuclear power industry in the late 1980s, no projections for China are included in the above tables because of lack of reliable data.

---

[1] Actual data.
[2] Estimated data for 1982.

Source: OECD Nuclear Energy Agency. *Nuclear Energy and Its Fuel Cycle: Prospects to 2025,* OECD, Paris, 1982.

TABLE 7.2 Projections of Installed Nuclear Capacity (end of the year stated; low–high; net GWe)

Data Source and Region	1980	1985	1990	2000
OECD Europe				
1969 Red Book (1968)	88–118	—	—	—
1970 Red Book (1970)	99	200	—	—
1973 Red Book (1972)	87	184	373	—
1975 Red Book (1975)	65–79	165–212	264–380	—
1978 Yellow Book (1977)	60	107–146	195–273	310–560
1979 Red Book (1979)	54–61	100–113	166–209	275–407
1982 Yellow Book (1980)	47	94–95	142–157	223–317
1983 Red Book (1983)	—	93	129	—
OECD America				
1969 Red Book (1968)	124–128	—	—	—
1970 Red Book (1970)	158	295	—	—
1973 Red Book (1972)	138	295	539	—
1975 Red Book (1975)	89	223	426	—
1978 Yellow Book (1977)	66	125–157	214–287	400–810
1979 Red Book (1979)	68–72	112–134	177–214	307–462
1982 Yellow Book (1980)	58–60	96–119	138–156	185–235
1983 Red Book (1983)	—	90	128	—
OECD Pacific				
1969 Red Book (1968)	16–18	—	—	—
1970 Red Book (1970)	24	60	—	—
1973 Red Book (1972)	33	63	106	—
1975 Red Book (1975)	17	49	85	—
1978 Yellow Book (1977)	15	27–40	50–80	140–270
1979 Red Book (1979)	17	26–33	45–60	100–151
1982 Yellow Book (1980)	15	28–30	51–53	89–131
1983 Red Book (1983)	—	26	46	—
OECD Total				
1969 Red Book (1968)	228–264	—	—	—
1970 Red Book (1970)	281	555	—	—

1973 Red Book (1972)	253	542	1018	—
1975 Red Book (1975)	171-185	437-484	774-890	—
1978 Yellow Book (1977)	141	259-343	459-640	850-1640
1979 Red Book (1979)	139-150	238-280	388-484	682-1020
1982 Yellow Book (1980)	121-123	218-244	330-366	497-683
1983 Red Book (1983)	—	209	303	450-470
Developing World				
1970 Red Book (1970)	23	55	—	—
1973 Red Book (1972)	11	25	50	—
1975 Red Book (1975)	9	46	114	—
1978 Yellow Book (1977)	5	19-25	45-60	150-250
1979 Red Book (1979)	5-9	19-23	45-50	150-186
1982 Yellow Book (1980)	3	14	30-33	88-121
1983 Red Book (1983)	—	13	21	54-88
WOCA				
1970 Red Book (1970)	300	610	—	—
1973 Red Book (1972)	264	567	1068	—
1975 Red Book (1975)	179-194	479-530	875-1004	—
1978 Yellow Book (1977)	146	273-368	504-700	1000-1890
1979 Red Book (1979)	144-159	257-303	434-534	832-1206
1982 Yellow Book (1980)	124-126	232-258	361-399	585-804
1983 Red Book (1983)	—	224	326	504-558
CMEA (+Yugoslavia)				
1979 Red Book (1979)	23-28	49-82	98-165	—
1982 Yellow Book (1980)	17	34-35	62-85	166-291
World				
1979 Red Book (1979)	167-187	306-385	532-699	
1982 Yellow Book (1980)	141-143	266-293	432-484	751-1095

Note about data sources: The first date is the publication date; the date between brackets is the date at which projections were made.

Sources: Adapted from OECD Nuclear Energy Agency. *Nuclear Energy and Its Fuel Cycle: Prospects to 2025,* OECD, Paris, 1982; and OECD Nuclear Energy Agency and the International Atomic Energy Agency. *Uranium Resources, Production and Demand,* OECD/IAEA, Paris, December 1983.

diagramatically in Figure 7.1. The enthusiasm with which many U.S. and Western European utilities (and national governments) in the early 1970s embraced the concept of nuclear power as a source of cheap and plentiful electricity is immediately apparent. As the decade progressed, however, early, overly optimistic projections of future levels of WOCA installed nuclear capacity were rapidly reduced until, by 1980, the projection for 1990 was only about one-third of the figure predicted ten years earlier. This trend has continued into the 1980s, with NUEXCO[3] estimating (in October 1983) that WOCA installed nuclear capacity would be only 281.5 GWe in 1990 (of which the United States would account for 108.3 GWe), a reduction of 14 percent on the estimate made by the OECD/IAEA[4] just ten months earlier.

If the data given in Table 7.2 are disaggregated to country level, a number of striking facts emerge. Since 1979 not only have there been no new orders for nuclear plants in the United States, but a considerable number of existing orders and partly constructed plants have been canceled or deferred. Over the period 1979 through 1983, 52 U.S. nuclear plants, representing a total capacity of about 57 GWe, were canceled at various stages of their construction. This figure is approximately 80 percent of U.S. installed nuclear capacity at year-end 1983. Thus even the projections given in the 1983 Red Book are at least 10 percent above a realistic estimate for the United States. Another example of a drastic curtailment of a nuclear-power program has occurred in Italy. In 1979, the OECD/IAEA Red Book estimate of installed nuclear capacity for Italy in 1990 was given as the range 25.9–33.4 GWe (from a 1979 level of 1.4 GWe). Just two years later the same publication revised its estimate for 1990 to only 7.4 GWe.

By way of contrast, ambitious plans for the rapid and continuing growth of nuclear power in both France and Japan have suffered little from the problems that have plagued their U.S. and Italian counterparts. In the case of Japan, this can be seen as an attempt to reduce its overdependence on oil-fired power stations which in 1980 still accounted for 46 percent of total electricity production. With few natural energy resources and limited hydro potential, Japan's reliance on nuclear power is expected to increase significantly by the year 2000, as illustrated in Table 7.3. A large expansion in electricity generated by coal-fired stations is also foreseen. A slower-than-anticipated rate of economic growth, however, has forced the Japanese government to reduce its nuclear-power expansion program in line with a corresponding reduction in the growth of demand for electricity. In mid-1983,

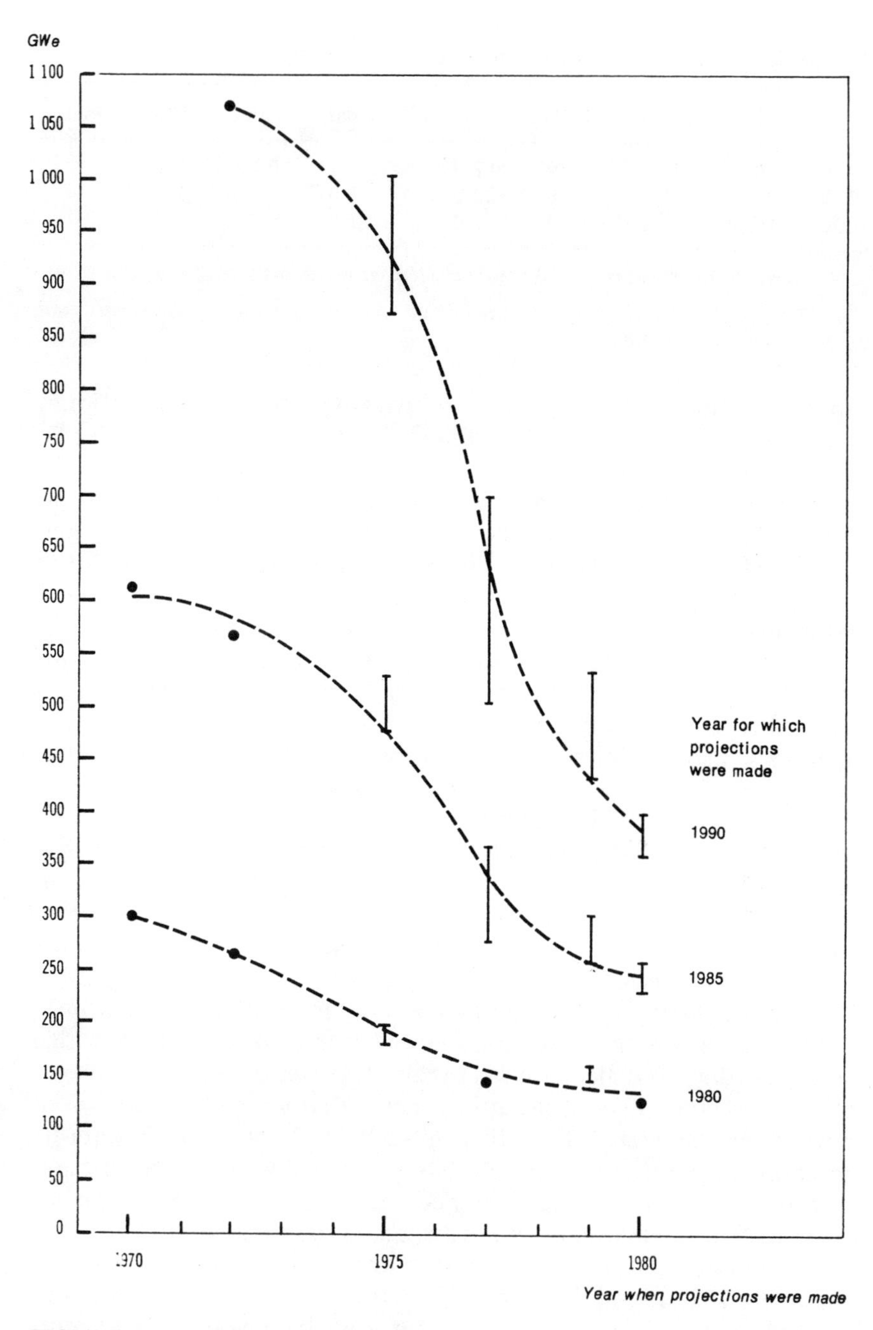

FIGURE 7.1 Estimates of Nuclear Capacities for WOCA (based on Table 7.2). *Source:* OECD Nuclear Energy Agency. *Nuclear Energy and Its Fuel Cycle: Prospects to 2025,* OECD, Paris, 1982.

TABLE 7.3 Japanese Electricity Production (Percentage)

	Oil	Hydro	LNG	Nuclear	Coal	LPG	Geothermal
1980	46.0	22.2	15.2	12.0	4.1	0.5	0.1
1990	22.5	21.1	20.6	22.0	11.0	1.6	1.3
2000	15.0*	20.0	16.0	30.0	17.0	*	2.0

*The 2000 estimate for LPG is included with the percentage for oil.

Source: Varley J.: "Update on Japan." *Nuclear Engineering International,* 28, No. 342 (June 1983).

Japan's nuclear-power capacity that was expected to be operational by year-end 1990 amounted to 35.3 GWe, a reduction of just over 30 percent on the range quoted earlier in the year by the OECD/IAEA and given in Table 7.1. The former figure corresponds to the total for nuclear reactors currently in operation, under construction, and already authorized. An installed nuclear capacity of 90 GWe is foreseen for the year 2000, but this figure appears to be extremely ambitious.

In 1982 nuclear power accounted for amost 40 percent of all electricity generated in France. With no plans for new coal-fired plants, France is placing a heavy reliance on an expanded nuclear capacity to provide its power requirements for the remainder of the century. France's nuclear capacity is expected to increase by 60 percent between 1982 and 1985, and to rise to about 60 GWe by 1990 (Table 7.1). With so much nuclear capacity in prospect and the continuing slow rate of growth in demand, the possibility of France's becoming an energy exporter on a large scale may arise during the 1980s. By 1983, France was exporting 5 percent of its electricity production.

The nuclear-power industry has long lead times, due not only to the large-scale nature of the construction project itself, but also the regulatory and licensing controls imposed by various governments. Consequently, the unforeseen decline in the growth of demand for electricity after 1974 coincided with additional generating capacity coming on-stream. Thus while few new construction projects for nuclear plants have been started since 1979, nevertheless installed nuclear capacity is projected to increase substantially during the decade of the 1980s as plants ordered during the first half of the 1970s (and delayed for various reasons) come on-stream. Thereafter, however, the momentum will disappear as few countries have follow-on construction programs for the 1990s. No new domestic orders have been placed in the Federal Republic of

Germany since 1975 or in the United States since 1979, while expansion plans in Italy, Spain, Sweden, and the United Kingdom have been severely restricted by drawn-out environmental and safety arguments, combined with reduced pressure for addditional electricity supply during the industrial recession.

The level of operational nuclear capacity at year-end 1983 is given in Table 7.4 on a country-by-country basis, together with the level of capacity that is currently under construction in each country. Under current conditions it is unlikely from a lead-time standpoint that a nuclear plant could be ordered, constructed, and licensed to operate all within the 1984–90 time frame. Thus, while projects included in the under construction category may still be subject to delay or cancellation (especially in the United States), combining these figures with operable capacity will yield a reasonable estimate of the maximum level of operable capacity in 1990. For OECD nations, therefore, a figure of 290 GWe for 1990 can be obtained by simply adding these two totals. This figure amounts to about 83 percent of the mid-range projection for OECD nations in 1990 given in Table 7.1. The corresponding value for the developing nations is 66 percent, while a rapid expansion of the nuclear power construction program in the USSR ensures that the centrally planned economies total for 1990 deviates only slightly from the mid-range forecast given in Table 7.1.

Table 7.4 also shows the percentage share of electricity that was supplied by nuclear power for each country in 1983. With such a large amount of capacity under construction the 1983 ranking of nations is likely to change substantially over the decade of the eighties. In that year, however, France placed the heaviest reliance (48.3 percent) on nuclear power for its electricity requirements, followed by Belgium (45.7 percent), Finland (41.65 percent), Sweden (36.9 percent), Bulgaria (32.3 percent), and Taiwan (31.3 percent). The average share of electricity requirements met by nuclear power in all OECD nations combined in 1983 was 16 percent. The IAEA estimate that comparable figures for 1990 and the year 2000 will be 23 and 25 percent, respectively.[5]

CHARACTERISTICS OF THE NUCLEAR-POWER INDUSTRY

Once a nuclear-power reactor is built it is virtually impossible to operate it on alternative fuels, other than uranium. Thus, over the

TABLE 7.4 World Nuclear Capacity (year-end 1983)

Country	Operable Capacity		Capacity under Construction		Electricity supplied by nuclear power reactors in 1983 (% of total)
	Units	GWe	Units	GWe	
OECD Nations					
Belgium	6	3.473	2	2.012	45.7
Canada	15	8.303	8	5.925	12.9
Finland	4	2.206			41.5
France	36	26.903	25	29.200	48.3
Germany (FRG)	16	11.110	11	11.908	17.8
Italy	3	1.232	3	1.999	3.2
Japan	28	19.023	10	10.022	20.0*
Netherlands	2	0.501			5.9
Spain	6	3.760	9	8.369	9.1
Sweden	10	7.355	2	2.100	36.9
Switzerland	4	1.940	1	0.942	29.3
United Kingdom	35	8.304	7	4.252	17.0
United States	80	63.315	50	55.738	12.7
total	245	157.425	128	132.467	n.a.
Developing Nations					
Argentina	2	0.935	1	0.692	8.8
Brazil	1	0.626	1	1.245	0.1
India	5	1.030	5	1.100	2.2*
Korea (South)	3	1.789	6	5.474	18.4
Pakistan	1	0.125			1.0
Taiwan	4	3.110	2	1.814	31.3

total	16	7.615	15	10.325	n.a.
Centrally Planned Economies					
Bulgaria	4	1.632	2	1.906	32.3
Czechoslovakia	2	0.762	9	4.354	8.0
Germany (GDR)	5	1.694			12.0*
Hungary	1	0.408	3	1.224	10.0
Soviet Union	43	20.671	41	38.001	8.0*
Yugoslavia	1	0.632			6.0*
total	56	25.799	55	45.485	n.a.
Others[1]			11	6.679	
World Total	317	190.839	209	194.956	12.0*

*Estimated by the IAEA.

[1] The "others" category contains 7 nations which have no operable nuclear capacity but which have capacity under construction. Details are as follows:

	Units	GWe (under construction)
China	1	0.300
Cuba	1	0.408
Mexico	2	1.308
Philippines	1	0.621
Poland	2	0.880
Romania	2	1.320
South Africa	2	1.842

Source: International Atomic Energy Agency Bulletin, 25 (December 1984): p. 71.

life of a reactor, the commitment to uranium is virtually irrespective of price; i.e., the demand for uranium is price inelastic. There are, however, a number of factors that could increase or decrease the demand over a period of time. The actual uranium requirements of the nuclear-power industry can be derived, given information concerning:

1. the characteristics of nuclear reactors in operation (i.e., the reactor mix);
2. the capacity utilization of nuclear-power plants (i.e., the load factor);
3. the tails assay at which the enrichment plants are operated;
4. the reuse of uranium and plutonium from spent reactor fuels.

We now consider these four factors individually.

Reactor Types

The majority of reactors in operation in the world today are Light Water Reactors (LWRs), which account for about 90 percent of installed nuclear generating capacity. This dominance is likely to last for a considerable time. The International Nuclear Fuel Cycle Evaluation (INFCE)[6] "low" forecast for installed nuclear capacity in the year 2000 gives LWRs about 87 percent of the total. Although the Heavy Water Reactor (HWR) is more economical than the LWR in its use of uranium, it will have little impact on the market during the remainder of this century.

While there is considerable potential for improving the efficiency of the LWR, the major long-term technological development in the nuclear-power industry is likely to be the introduction of the Liquid Metal Fast Breeder Reactor (LMFBR). Although the economics of electricity generation using fast-breeder reactors are uncertain at present,[7] their fuel costs are extremely low and their widespread introduction would serve to place an upper limit upon the amount of new uranium that may eventually be required. The widespread introduction of the LMFBR could ultimately eliminate the need for mining uranium, since LMFBRs could use plutonium combined with depleted uranium obtained from the stockpile generated by the enrichment process.

Up until the mid-1970s, it was widely assumed that the LMFBR would be introduced commercially on a substantial scale before the end of this century and, when adopted, their effect on the market for uranium would be dramatic. In 1976 the OECD produced a "high" projection for the level of installation of the LMFBR of almost 10 percent of installed WOCA nuclear capacity by the year 2000. Just three years later the same source gave a "high" projection of about 3 percent for the same year. This substantial reduction was largely brought about by the U.S. government's decision in 1977 to defer the introduction of fast-breeder reactors for commercial use. It now appears that the LMFBR will not play a substantial role in electric-power generation until well into the next century.

Power Plant Load Factors

Nuclear-power plant utilization can be an important factor in the demand for uranium. The average capacity utilized in a particular year expressed as a percentage of the design capacity is known as the load factor. In the past, utilities have tended to overestimate the load factor for nuclear plants due to unplanned maintenance and retrofitting to meet new regulation standards. The effect of this miscalculation has been to decrease uranium requirements in the short run as reactors required refueling less frequently than originally estimated. Of particular note in this context is the aftermath of the Three Mile Island accident in 1979. During 1978 the load factor for all U.S. nuclear-power reactors averaged 63.9 percent. The following year it fell to 57.6 percent and then to 53.8 percent in 1980. The trend was reversed in 1981 when the U.S. load factor rose to 57.4 percent, but since then it has remained relatively constant. Table 7.5 gives 1983 average annual load factors for nations possessing more than four nuclear reactors. It is interesting to note that two large-scale users of nuclear power—namely, France and the United States—fall very low on the list. This is particularly relevant when it is remembered that OECD/IAEA projections of WOCA uranium requirements are based upon an assumed load factor of 70 percent.

The Operation of Enrichment Plants

The way that enrichment plants are operated is an important determinant of the demand for uranium. Current techniques allow a tails assay ranging from 0.15 percent to 0.30 percent. If laser enrichment were found to be a commercially viable proposition,

TABLE 7.5 Average Annual Load Factors (1983)

Country	Average Annual Load Factor (%)
Switzerland	88.0
Finland	86.4
Canada	78.1
Belgium	74.6
Japan	70.0
Germany (FRG)	69.8
Taiwan	69.8
Sweden	63.4
France	61.8
United Kingdom	59.3
United States	54.5
Spain	44.8
India	38.7

Source: Laurie Howles, "Nuclear Station Achievement." *Nuclear Engineering International,* 29, No. 355 (May 1984): pp.36–38.

this could be reduced to 0.01 percent. It should be remembered, however, that the optimal tails assay is determined by both the cost of uranium *and* the cost of enrichment services. The economics of uranium enrichment is discussed in detail in Chapter 10.

Spent Fuel Reprocessing and Recycling

Spent fuels can be reprocessed to provide uranium and plutonium, or uranium alone. The uranium is only slightly more enriched than natural uranium, and consequently it affects the market for uranium and not the market for enrichment services. Plutonium, however, both replaces uranium and decreases the number of separative work units required in the enrichment process.

The impact of recycling on uranium requirements depends on whether both uranium and plutonium are recovered and on the rate at which recovery facilities are built. At a maximum, full reuse of both metals would ultimately decrease uranium requirements by about 30 percent, while recycled uranium alone would decrease uranium requirements by about 19 percent. Thus policies regarding recycling combined with policies on tails assay can substantially change uranium requirements for any given nuclear-power forecast.

Commercial reprocessing only exists in a few countries at present (although many more countries operate pilot plants), and the economic viability of such projects is uncertain. A comprehensive analysis of the comparative economics of recycling and reprocessing has been carried out by the U.S. Nuclear Regulatory Commission.[8] The net benefit they predicted would accrue as the result of recycling and reprocessing over the period 1976–2000 amounted to a reduction of less than 1 percent in the price of electricity. This was very small when compared with the uncertainties in the cost estimates on which the analysis was based.

Projections for the gradual introduction of recycling and reprocessing have been moved further into the future which, other things being equal, has increased the requirements for uranium. Delays in the introduction of recycling are partly due to environmental and regulatory problems, but the current oversupply of uranium at "low" prices probably renders recycling uneconomic at present (and for the foreseeable future). The future of plutonium recovery is particularly uncertain, especially in the United States. In addition to safety and transportation problems, the specter of a "plutonium economy" led the United States in 1977 to defer indefinitely the reprocessing and recycling of plutonium.

REACTOR REQUIREMENTS

Once a 1-GWe light water nuclear-power reactor is built, its lifetime fuel requirements average about 5,500 tons U_3O_8, assuming a 70 percent load factor and a 0.20 percent tails assay at the enrichment plants. A more detailed analysis of the fuel requirements for different types of reactors is given in Table 7.6. Over the remainder of this century, the Light Water Reactor, and in particular the Pressurized Water Reactor, will dominate the nuclear-power industry. The table indicates, however, that significant fuel savings can be made with improved versions of the PWR that are expected to enter service over the next 20 years. If Fast Breeder Reactors are introduced on a substantial scale over future years it is apparent from the table that ultimately requirements for new uranium will diminish. Such a possibility, however, appears remote at present.

The lead times for the construction and licensing of nuclear-power plants have always been long. Largely because of increased time periods for regulatory approval, they have lengthened in the United States from about six years in the 1960s to about 12–15 years today. Consequently, the utilities have considerable leeway in the timing of their orders for the initial core. Their purchases for

TABLE 7.6 Lifetime and Annual Requirements for a 1 GWe Nuclear Power Reactor for Various Fuel Cycles (70% load factor; 0.20% enrichment tails)

Reactor Type	30-Year Life (tons U_3O_8)	Annual Equilibrium (tons U_3O_8/annum)
Once-through reactors		
LWR (current technology)	5,500	174
PWR (15% improved)	4,800	152
PWR (30% improved)	4,000	122
AGR (updated)	6,100	190
GG-MAGNOX	9,300	273
HWR-CANDU (current)	4,800	156
HWR-CANDU (improved)	3,400	107
Recycle reactors		
(Conversion ratio less than 1; 1 year recycle delay)		
LWR-PU recycle	4,000	118
HTR-U recycle, HEU makeup	2,100	46
HWR-CANDU, Nat. U + Pu recycle	2,700	77
HWR-CANDU, thorium, U recycle and HEU makeup	1,900	33
HWR-CANDU, thorium, U recycle and Pu makeup	2,300	52

Recycle reactors[1]	Tons U_3O_8 Required to Establish 1 GWe
(Conversion ratio greater than or equal to 1; 1 year recycle delay)	
Reference FBR—Pu from LWR	5,000
" " —Pu from 15% improved PWR	6,000
" " —Pu from AGR (updated)	8,200
" " —Pu from MAGNOX	3,000
" " —Pu from CANDU (current)	2,200
" " —Pu from CANDU (improved)	3,500
HWR-CANDU, self-sufficient thorium cycle initiated by HEU	1,100
HWR-CANDU, self-sufficient thorium cycle initiated by Pu from Nat. U	2,200

Note: HEU = Highly Enriched Uranium

[1]These reactors or their replacement units will operate indefinitely without further requirements of natural uranium. However, the self-sufficient thorium cycle (conversion ratio 1) requires additional uranium, equal to the amount shown above, for each "new" unit added to the system. The FBR, by contrast, produces an excess of Pu such that it can supply the required total inventory for another FBR unit after 19 or 26 years of operation according as the recycle delay is one or two years.

Source: Adapted from OECD Nuclear Energy Agency, *Nuclear Energy and Its Fuel Cycle: Prospects to 2025*, OECD, Paris, 1982.

refueling can also be advanced or postponed, depending on their inventories, current prices and expected price movements, and their contract schedule with the enrichment plants. While there is no direct substitute for uranium in the short run (unless the utilities shift the loads toward their fossil fuel plants), the demand for uranium can be very volatile because of the timing of reactor orders, cancellations, and deferrals of nuclear-power units and because of the options for scheduling purchases of uranium. Long lead times and the ability of utilities to defer purchases contribute to substantial uncertainties in supply decisions, particularly concerning exploration and development of mines.

DEMAND AND CONSUMPTION

The demand for uranium in any time period (D_t) can be defined as the sum of the perceived level of consumption in that period (C_t^*) and the planned change in the level of inventories over the preceding period (ΔI_t).

Algebraically, this relationship can be expressed as:

(a) $D_t = C_t^* + \Delta I_t$

where $\Delta I_t = I_t - I_{t-1}$

Consumption is defined as the amount of uranium entering the conversion–enrichment–fuel fabrication pipeline. Thus consumption should largely be determined by the current (and short-term future) level of installed nuclear-generating capacity. The rigid nature of U.S. enrichment contracts that were in force during the second half of the 1970s, however, combined with the recent experience of reactor deferrals, cancellations, and delays in licensing, has meant that "consumption" has been higher than warranted by the level of nuclear-generating capacity alone. This has been reflected in a buildup of inventories of enriched uranium. Thus "actual consumption" as defined here will differ from reactor consumption, the difference being accounted for by variations in the level of inventories at the conversion, enrichment, and fuel-fabrication stages.

Since the perceived level of consumption may not be realized due to unplanned maintenance, plant breakdowns, licensing delays, etc., we can express this variable as the sum of two components, namely:

(b) $C_t^* = C_t + S_t$

where C_t represents the actual level of consumption and S_t, a residual term, can be regarded as unplanned changes in the level of inventories. Clearly S_t can be negative if actual consumption exceeds perceived consumption, but a number of events that occurred during the 1970s exerted a major influence on the uranium market (namely, a major industrial recession, the accident at Three Mile Island in 1979, and reactor licensing delays) and resulted in a lower level of consumption than was envisaged in earlier years, thus resulting in positive values for S_t.

Upon substitution for C_t^* from equation (b) into equation (a) we obtain

(c) $D_t = C_t + \Delta IS_t$

where $\Delta IS_t = \Delta I_t + S_t$ is the change in the sum of planned and unplanned inventories over the preceding period. While ΔI_t is largely within the control of uranium consumers, by definition S_t will be an unknown quantity.

Forecasts of "apparent future consumption" of uranium, calculated on a reactor-by-reactor basis by NUEXCO, are given in Table 7.7. These projections refer to requirements for reactors currently operating, under construction, or on firm order, and thus ignore reactors that may come on-line by 1990 but have yet to be ordered. They also ignore contractual obligations which utilities may have with suppliers of conversion, enrichment, or fuel fabrication services in excess of current reactor requirements. The actual level of future uranium demand, however, may exceed or fall short of apparent future consumption, depending upon the consumers' inventory policies.

WOCA uranium consumption is projected to increase by almost 40 percent between 1983 and 1990, with the bulk of this expansion taking place in France, Japan, and the United States. This relatively rapid rate of growth reflects the coming on-stream of additional nuclear capacity in France and Japan, which was ordered during their expansionary phases in the late 1970s. In the case of the United States, longer lead times mean that many plants ordered in the early 1970s will be entering service during the 1980s. Comparable estimates compiled by the OECD/IAEA[9] are

TABLE 7.7 Apparent Future Consumption (thousand tons U_3O_8)

Buyer Group	1983	1984	1985	1986	1987	1988	1989	1990	1991	1992
United States	14.5	14.8	15.3	18.2	19.3	20.0	19.5	18.8	19.7	19.2
Europe	17.2	18.3	19.2	20.1	19.4	20.8	21.9	23.8	22.8	25.8
Far East	6.0	6.2	6.9	6.8	7.3	7.4	9.2	9.0	10.7	9.6
Other	2.2	2.2	2.3	2.8	2.7	2.9	3.7	3.7	3.9	3.3
total WOCA[1]	39.8	41.4	43.6	47.9	48.6	50.9	54.2	55.2	57.1	57.9

[1] Individual figures do not sum to the total due to rounding.

Source: NUEXCO, Monthly Report on the Nuclear Fuel Market, October 1983.

50.5 and 71.4 thousand tons U_3O_8 for 1985 and 1990, respectively. The difference between these two sets of estimates is largely due to the OECD/IAEA's overestimate of U.S. installed nuclear capacity to 1990.

Forecasting future levels of uranium consumption is a complex exercise involving projections of electricity requirements, technological change, and relative fuel costs in order to estimate the role that nuclear power will play over forthcoming years. When projections of future levels of installed nuclear capacity have been obtained, subsequent projections of uranium consumption will depend upon the optimal tails assay, estimated load factors, and improvements in both reactor and fuel efficiency. The potential for inaccuracy, therefore, is great, as was noted earlier with respect to estimates of future levels of installed nuclear capacity made during the 1970s.

Between 1965 and 1983, WOCA uranium production totaled almost 650,000 tons U_3O_8. During the same period reactor requirements totaled less than 400,000 tons. Much of the difference between these two amounts is held as inventory by consumers or producers. In addition, some of the pre-1965 production (320,000 tons U_3O_8), not used for military purposes, could also be included in inventories. Few countries, however, provide detailed information on the size of their uranium inventory, while the inventory-holding policies of the major uranium-consuming nations are very variable, ranging from one to four years of forward consumption.

A striking feature of the commercial uranium market is that, since its establishment in 1968, world production has *always* exceeded consumption, often by a factor of more than two! By comparing Tables 6.2 and 7.7, however, it is apparent that this situation is coming to an end, and by the late 1980s consumption and production should be in equilibrium. Of particular importance is the level of relative inventory, which reflects the number of years forward consumption that can be supplied by the current level of total inventory. Table 7.8 summarizes NUEXCO's projections of changing inventory levels over the years to 1992. The level of relative inventory is expected to decrease gradually during the remainder of the 1980s, falling from 4.6 years in 1983 to 3.9 years by 1990. The relative level of inventory in the United States, however, is expected to decline far more rapidly, from 4.2 years in 1983 to 1.5 years by 1990 (Table 7.9). NUEXCO believes this latter figure to be the collective minimum level of inventory that U.S. utilities would expect to hold to satisfy immediate and short-term requirements. This belief is supported by a Department of Energy

TABLE 7.8 Uranium Inventory: Total WOCA (thousand tons U_3O_8)

	1983	1984	1985	1986	1987	1988	1989	1990	1991	1992
Net inventory increase (decrease)	7.55	6.10	8.25	9.10	6.60	1.80	(5.70)	(3.60)	(2.60)	(2.70)
Total inventory[1] (year-end)	212.00	218.10	226.35	235.45	242.05	243.85	238.15	230.45	227.85	225.15
Relative inventory[2] (years)	4.6	4.5	4.4	4.4	4.3	4.2	4.1	3.9	3.8	3.6

[1] This figure assumes a beginning inventory of 204.45 thousand tons U_3O_8.
[2] Relative inventory is the number of years' forward consumption which could be supplied by total inventory at year-end of the previous year.

Source: As for Table 7.7

TABLE 7.9 Uranium Inventory: United States (thousand tons U_3O_8)

	1983	1984	1985	1986	1987	1988	1989	1990	1991	1992
Net inventory increase (decrease)	(3.30)	(6.45)	(4.65)	(6.95)	(8.00)	(7.05)	(2.80)	0.35	0.60	0.25
Total inventory[1] (year-end)	63.90	57.45	52.80	45.85	37.85	30.80	28.0	28.35	28.95	29.20
Relative inventory[2] (years)	4.2	3.8	3.2	2.8	2.3	1.9	1.5	1.5	1.5	1.5

[1] This figure assumes a beginning inventory of 67.20 thousand tons U_3O_8.
[2] NUEXCO assume that U.S. utilities will maintain a collective minimum of 1.5 years' forward requirements in inventory.

Source: As for Table 7.7.

survey in 1983 which revealed that a majority of U.S. utilities with specific policies regarding inventories considered a period of up to two years of forward coverage for U_3O_8 as a desirable level of inventory.[10]

CONCLUDING REMARKS

This chapter has sought to identify and discuss the major determinants of uranium demand. It is clear that the current and projected excessive levels of WOCA uranium inventories have been brought about by a slump in demand caused by a number of factors—namely: an industrial recession, delays in licensing nuclear reactors, public concern regarding safety aspects of nuclear power, and overly optimistic forecasts made in the early 1970s.

At the same time, WOCA uranium production has been slow to respond to changes in market conditions. Many high-cost producers have been spared an early demise by import quotas (United States) and favored-nation trading policies (in particular, France with its former colonies of Gabon and Niger). As a consequence, uranium production during the late 1970s/early 1980s was approximately double consumption, with the surplus passing into a largely unintended buildup of inventories. The fairly rapid expansion of uranium requirements that is forecast for the 1980s should ensure that this surplus diminishes over the decade, but excessive levels of inventories are certain to maintain a dampening influence on the price of uranium until the turn of the decade unless a major stimulation of demand can be achieved.

While an economic recovery in industrialized nations during the mid-1980s would be a prerequisite for stimulating uranium demand, public and government attitudes toward nuclear power must undergo a radical change (especially in the United States) to allow for more streamlined regulations, and consequently fewer costly delays, in the construction and licensing of nuclear-power reactors. A 1983 estimate of the economic viability of U.S. nuclear-power plants scheduled to begin operations in 1995 showed that, in many areas of the United States, their cost per GWe would be slightly higher than that of comparable coal-fired plants,[11] although their European counterparts still appear to have significant cost advantages over coal.[12] If the leadtime between construction and commercial operation of U.S. nuclear-power plants could be significantly reduced, then power utilities may regard them, once again, as a feasible economic alternative to the economically (and politically) more acceptable coal-fired plants.

NOTES

1. Eklund (1981).
2. *NUEXCO, Monthly Report on the Nuclear Fuel Market*, October 1983.
3. *Ibid.*
4. OECD/IAEA (1983).
5. *International Atomic Energy Agency Bulletin*, No. 25, p. 70, December 1984.
6. Reported in OECD/IAEA (1979).
7. Chow (1980) evaluates the economics of the breeder reactor vis-à-vis its light water competitor.
8. Reported in Nuclear Energy Policy Study Group (1977).
9. OECD/IAEA (1983).
10. *Survey of United States Uranium Marketing Activity*, Energy Information Administration, Department of Energy, Washington, D.C., 1982.
11. U.S. Department of Energy (1982).
12. OECD (1983).

8

Uranium Price Formation

INTRODUCTION

Following the establishment of a private market for uranium in the United States in the late 1960s, the ebb and flow of the industry's fortunes has been reflected by the price of uranium as represented by NUEXCO's exchange value. The latter value is a judgment of the price at which transactions for significant quantities of natural uranium concentrate could be concluded as of the last day of the month. NUEXCO's method of calculation, however, has changed over time, reflecting changing market conditions.[1] From August 1968 to October 1973 (buyer's market) exchange values were based upon current offers to sell; from November 1973 to May 1978 (seller's market) they were based upon bids to buy, and after May 1978 NUEXCO incorporated buyers' bids, sellers' offers, and transactions. Exchange values therefore are current prices for current or near-term delivery. While NUEXCO emphasizes that their exchange value is not a spot price in the usual sense of the word, nevertheless it is generally regarded as an indicator of uranium spot, or short-term, market-price levels.

The short-term market, however, only accounts for about 10–15 percent of uranium transactions. The bulk of WOCA's uranium requirements are procured under long-term contracts, which generally entail the consideration of different criteria for determining the price of uranium. The most accessible source of data on U.S. contract prices is the annual Survey of United States

Uranium Marketing Activity.[2] For deliveries of natural uranium made to member nations of the European Economic Community (EEC), quantities together with average prices are published annually in the Euratom Supply Agency's annual report. Comparable data for uranium deliveries to WOCA's third major source of uranium demand, Japan, are not made publicly available.

In the next section of this chapter, the two basic methods of contract-price determination are discussed and the variability of contract prices for delivery during the 1980s is illustrated. The remainder of the chapter discusses the major determinants of short-term uranium price formation as reflected by the exchange value. Although the influence of government policy and regulation changes, speculation, panic, and lack of information (or knowledge) on the short-term market should not be underemphasized, this study pays particular attention to the role of inventories in short-term price determination. Of necessity, the analysis in this chapter is mainly restricted to the U.S. uranium market because of the dearth of information concerning contracts and prices elsewhere in the world.

PRICES

There are two basic types of contract-price determination: contract price and market price.[3] Contract prices relate to procurement arrangements when prices and their escalation factors (if any) are determined at the time the contract is signed. In market-price contracts, however, the prices are typically (but not always) determined at or some time prior to delivery and are based on prevailing market prices. Most market-price contracts contain floor (i.e., minimum) prices that provide a lower limit on the eventual settled price. These floor prices and the means of escalation are generally determined when the contract is signed.

Contract Prices

Table 8.1 shows the average contract prices of 1982–91 deliveries for price data reported in the January 1983 survey. These prices are stated in year-of-delivery dollars and, as such, reflect the respondents' estimates of future values of the escalation factors in the contracts. Market-price settlements for 1982–83 are included with contract prices since, as settled prices, they are similar to contract prices. This procedure gives a generally comprehensive average price for actual deliveries in 1982. Also shown in Table 8.1 is an adjusted price, which includes estimates for prices not reported by the respondents to the survey. The adjusted price is a weighted average of reported prices and price estimates for contracts not supplying price information.

The average reported price for U.S.-origin uranium deliveries in 1982 was \$38.37/lb U_3O_8, an increase of \$6.17 over 1981. A small decline is envisaged to occur in 1983. Prices paid by U.S. buyers for imported uranium in 1982 and 1983 were \$25.19/lb and \$26.42/lb, respectively.

Figure 8.1 displays the distribution of reported U.S. contract prices and market-price settlements by year of delivery. For example, the price for 1982 deliveries ranged from approximately \$10/lb U_3O_8 to approximately \$70/lb. This figure illustrates the high degree of variability with respect to projected prices for future deliveries. This can largely be attributed to the types of escalation factors specified in the contracts, respondents' projections of the

TABLE 8.1 Average U.S. Contract Prices and Market-Price Settlements as of January 1, 1983 (year-of-delivery dollars)

Year	Reported Price per Pound of U_3O_8	Percent of Delivery Quantities with Reported Prices	Adjusted Price[1] per Pound of U_3O_8
1982	38.37[2]	87	39.82[2]
1983	35.62[2]	90	37.57[2]
1984	44.84	89	45.70
1985	50.00	81	49.66
1986	48.98	77	48.17
1987	52.20	71	51.14
1988	46.65	57	48.74
1989	49.59	65	52.21
1990	57.16	59	59.13
1991	58.91	64	61.90

[1] Adjusted price per pound includes estimates for prices not reported by the respondents.

[2] Includes settlements of market-price contracts. The 1982 and 1983 data do not include uranium delivered or to be delivered under litigation settlements.

Source: 1982 Survey of United States Uranium Marketing Activity, Energy Information Administration, U.S. Department of Energy, Washington D.C., September 1983.

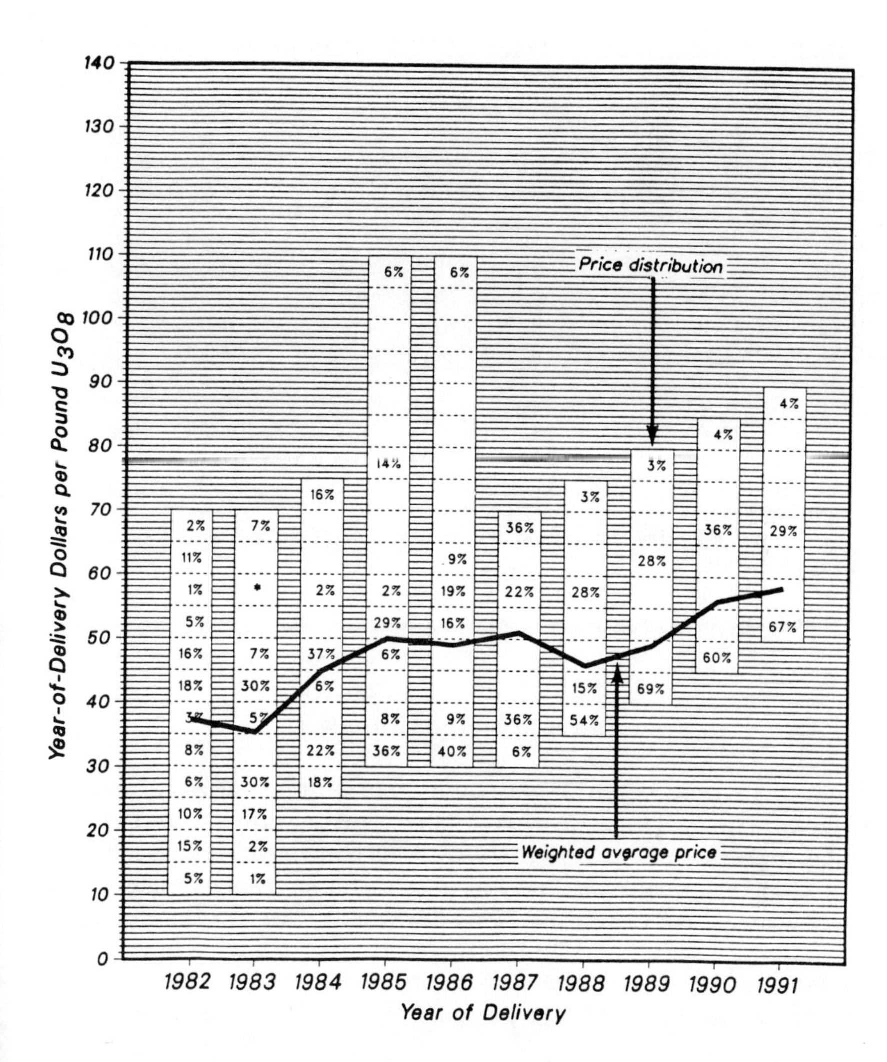

FIGURE 8.1 Distribution of Reported U.S. "Contract" Prices and Market-price Settlements as of January 1, 1983 (year-of-delivery dollars).
Source: As for Table 8.1.

*A blank cell indicates 0 percent.

value of these escalation factors, and the year the contracts were signed.

For deliveries made to the electrical utilities of the EEC during 1982 under medium- or long-term contracts, the average price (weighted by quantity) was \$32/lb U_3O_8, a decline of \$1.25 over the previous year.

TABLE 8.2 Average Floor Prices of U.S. Market-Price Contracts as of January 1, 1983 (year-of-delivery dollars)

Year	Reported Price per Pound of U_3O_8	Percent of Commitments with Reported Floor Prices	Adjusted Price[1] per Pound of U_3O_8
1982	51.27	74	52.20
1983	53.49	93	54.04
1984	55.93	94	56.61
1985	61.05	92	61.93
1986	62.84	91	64.18
1987	65.50	90	67.34
1988	70.74	100	70.74[2]
1989	75.05	100	75.05[2]
1990	72.39	100	72.39[2]
1991	76.85	100	76.85[2]

[1] Adjusted price per pound includes estimates for prices not reported by the respondents.

[2] Estimates of missing prices not required.

Source: As for Table 8.1.

Market Price Contracts

Table 8.2 summarizes the average reported floor prices of market-price contracts. These average reported prices range from \$51.27/lb U_3O_8 in 1982 to \$76.85/lb in 1991. Also shown in Table 8.2 are the adjusted floor prices. Table 8.3 shows the average settlements of market-price contracts for 1982 deliveries. The average price for settlements with and without price floors was \$41.05/lb, an increase of \$2.85 over the corresponding figure for 1981. Table 8.3 also shows that the average price for 1982 settlements with no price

TABLE 8.3 Average Reported U.S. Price Settlements of Market-Price Contracts for 1982 Deliveries

Dollars per Pound of U_3O_8	Millions of Pounds of U_3O_8
21.50 (no price floor)	2.8
51.27 (price floor only)	5.3
41.05 (with and without price floor)	8.1

Source: As for Table 8.1.

floor (i.e., the agreed-upon market price) was $21.50/lb, marginally higher than the average unweighted NUEXCO exchange value of $19.90/lb for the same year.

Figure 8.2 shows the annual ranges and averages of reported floor prices in market-price contracts during the 1983–91 period.

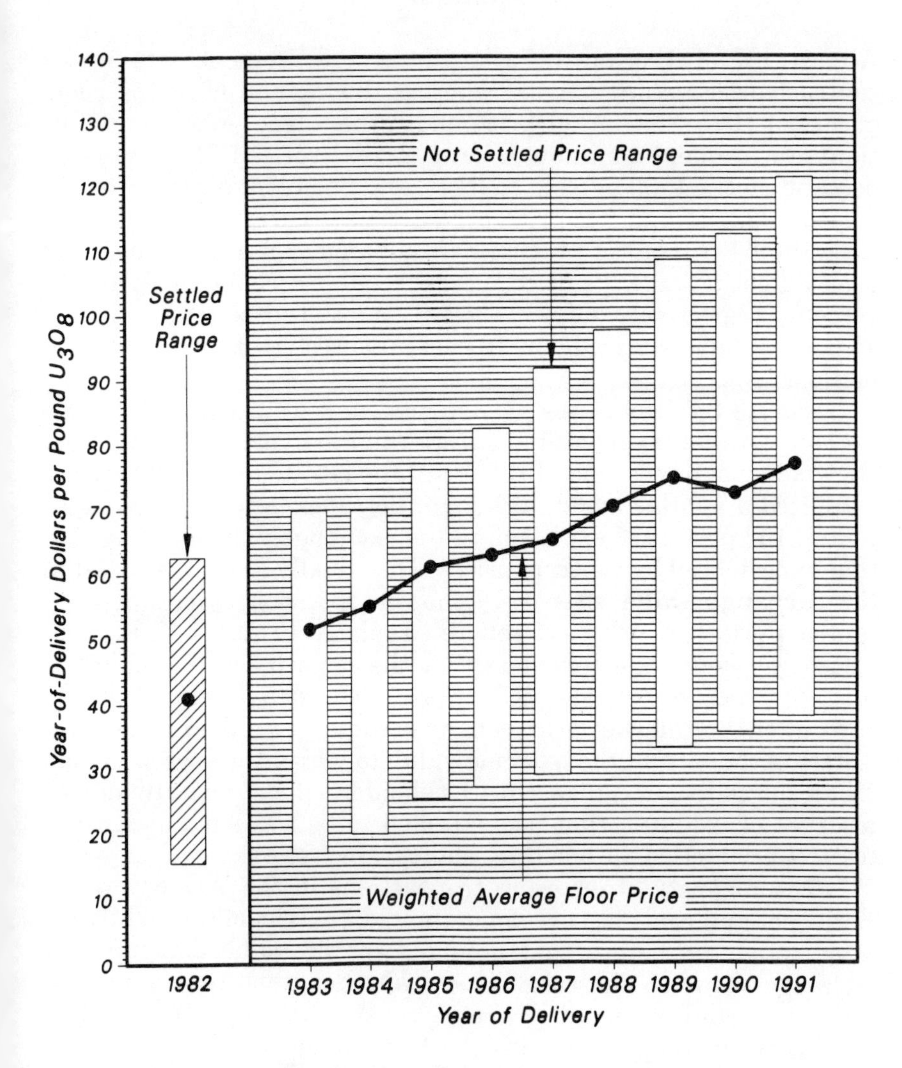

FIGURE 8.2 Range of Reported Floor Prices in U.S. Market-price Contracts as of January 1, 1983 (year-of-delivery dollars).
Source: As for Table 8.1.

Data for 1982 represent the range of market-price settlements. The variability of these figures is quite astonishing. The least variability is shown by the 1982 settled price range, but even here the floor prices range from about $16/lb U_3O_8 up to almost $63/lb.

Summary

Almost all of the pre-1975 uranium procurement was under contract price arrangements, whereas during and after the boom period of 1975–80 market price and "other" (e.g., captive production) procurement dominated. Uranium deliveries in 1982 were 42 percent contract, 53 percent market, and 5 percent other. Over the period 1982–91, market-price contracting is currently expected to make up 64 percent of procurement arrangements, while contract price and other arrangements are expected to account for 25 and 11 percent, respectively. Of the 1982–91 market-price commitments, 45 percent have a price floor, 18 percent a cost floor, and 37 percent no floor.

Figure 8.3 illustrates movements (in real terms) in both the exchange value (expressed as an unweighted yearly average) and DOE contract prices since 1971. Until the late-1970s, the two series exhibited a similar trend, although, not unexpectedly, the amplitude of the post-1973 explosion in the exchange value far exceeded that experienced by contract prices. The equally as dramatic fall in the exchange value after 1977, however, was not monitored by contract prices, which have remained relatively stable. This is not surprising when one considers that the majority of U.S. market-price contracts are subject to floor prices. For deliveries in 1982, 66 percent of U.S. market-price contracts contained a floor price, thus isolating them from the extreme value to which the spot price had fallen. In addition, contract prices should depend critically on the reserves of uranium that are expected to be available at various costs of production rather than short-term factors.

The majority of uranium transactions in the Western world (both spot and contract) are denominated in U.S. dollars. While we have noted the slump that occurred in the dollar-denominated exchange value in the early 1980s, another important factor for non-U.S. uranium producers and consumers is their domestic currency's exchange rate with the U.S. dollar. The early years of the 1980s have witnessed rapid changes in exchange rates, most notable of which was an appreciation of the trade-weighted value of the U.S. dollar in excess of 40 percent between January 1980 and

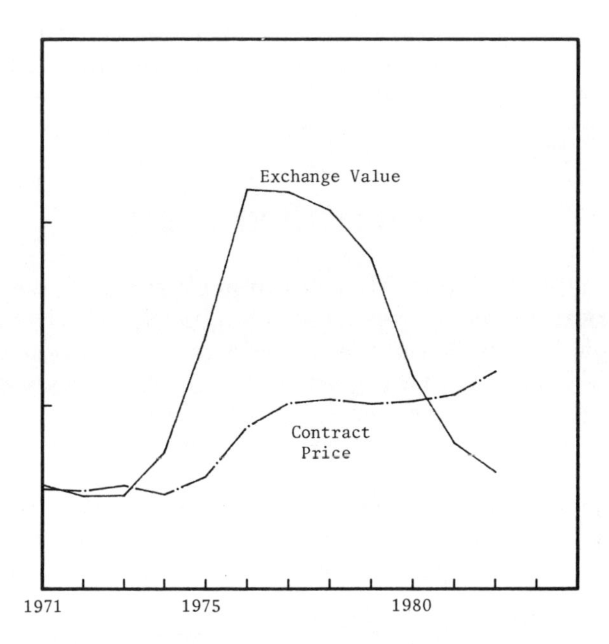

FIGURE 8.3 Uranium Short-term and Contract Prices (1967 prices).
Source: Average contract price: *Survey of United States Uranium Marketing Activity,* U.S. Department of Energy, various issues. Exchange value: NUEXCO reports.

The U.S. Producer Price Index for Industrial Commodities was used to deflate the two series.

year-end 1983. As a result of this (and other) major exchange rate "realignments," some countries were trading uranium on the near-term market in the early 1980s at domestic currency prices in excess of those prevailing at the height of the uranium boom of the late 1970s. By August 1983, the exchange value expressed in French francs or Swedish kronor was 19 percent and 14 percent, respectively, *above* its value at the beginning of 1980, whereas its U.S. dollar value was just 60 percent of that level. In general, the appreciation of the U.S. dollar had the effect of increasing unit revenue (i.e., export revenue per pound of U_3O_8) for uranium exporters while at the same time increasing unit costs (i.e., import payments per pound of U_3O_8) for uranium importers. The benefits of higher unit revenue for uranium-exporting nations, however, may well be offset by increases in the U.S.-denominated price of imported raw materials (especially for resource-poor nations like Niger) necessary to operate the uranium mines. In addition, non-U.S. mining projects that are financed by dollar-denominated loans will incur a higher level (in domestic currency) of repayment and debt-servicing costs.

URANIUM INVENTORIES

It is reasonable to assume that uranium consumers have a planned level of uranium inventories (at least implicitly) which they wish to hold to enable immediate and short-term requirements to be safeguarded. This can be regarded as a transactionary motive for holding inventories. In general, one would expect the level of such inventories to be directly related to the level of current consumption.

A further motive for holding inventories will be the consumer's perception of short-term uranium price movements. If the price of uranium is expected to rise in the near future then a utility may wish to increase its current level of inventories with a view to securing future supplies at the prevailing price. The subsequent gain (or profit) arising from such an action is simply the difference between the current price and the price prevailing at the date the uranium is required, less the cost of storage (i.e., the opportunity cost of holding the commodity as uranium rather than investing the money in government bonds, in addition to actual physical storage costs). We do not assume that the utility will subsequently sell the uranium to realize actually a profit, rather it may use the uranium itself at some later date. Nevertheless, it will still realize a speculative gain (or loss!). The converse will also hold true. If the price of uranium is expected to fall, and provided that industry inventories are adequate with no foreseeable threat to the availability of uranium, then utilities would be reluctant to purchase additional (to current requirements) quantities of uranium. They may even seek to run down the level of inventories if it is considered to be excessive. A similar result would occur if the real rate of interest were to increase, thus rendering some inventories too costly to hold.

The *planned* level of consumer inventories, therefore, will largely be determined by three factors: the current level of inventories relative to consumption, expected price movements, and a rate of interest. These three factors account for the transactionary motive, the speculative motive, and the cost of holding inventories, respectively. All three of these factors are, of

TABLE 8.4 United States Uranium Inventories, 1980–83 (short tons U_3O_8 equivalent)

	1/1/80	1/1/81	1/1/82	1/1/83
All Buyers				
Natural Uranium[1]	36,100[2]	45,100[2]	51,200[2]	57,550
Enriched Uranium	16,200	18,300	11,200	9,400
Fabricated fuel	*	*	10,200	9,900
Producers				
Natural Uranium[1]	2,400	2,700	7,000	13,300
Department of Energy				
Natural Uranium[1]	n/a	20,500	20,500	20,500
Enriched Uranium	n/a	61,400	59,200	58,100
total	52,300	148,000	159,300	168,750

*Included in enriched uranium figures.

[1] Includes U_3O_8 and natural UF_6.

[2] Does not include natural UF_6 inventories at DOE enrichment plants.

Source: Survey of United States Uranium Marketing Activity, U.S. Department of Energy, 1980–83.

course, conditional upon utility perceptions about the future physical availability of uranium.

It is also important to consider *unplanned* changes in the level of inventories. Reactor deferrals and cancellations have, over recent years, led to an accumulation of inventories by utilities which has been largely unintentional or inadvertent. We shall be concerned with total inventories, i.e., the sum of both planned and unplanned inventories held at all stages of the front end of the nuclear-fuel cycle.

Data relating to privately owned U.S. uranium inventories have been published by the Department of Energy, and previously ERDA, since 1970. This series, however, was subject to major definitional changes in 1972, 1979, and 1982. Originally the series reported inventories of U_3O_8 alone; then, beginning in 1972, inventory data included stocks of UF_6. The series was further expanded in 1979 to include inventories of enriched uranium, and again in 1982 to include inventories of fabricated fuel. Details of uranium inventories held by the Department of Energy were first published in 1981. Table 8.4 shows the composition of total U.S.

uranium inventories as of January 1 for the years 1980–83. For the latter year, the total inventory was 168,750 tons U_3O_8 equivalent, approximately half of which was owned by the Department of Energy. This figure represents the equivalent of more than nine years of forward requirements. With respect to privately owned inventories, buyer inventories of natural uranium is by far the most important category, accounting for 64 percent of all privately owned uranium inventories in 1983. Since this also happens to be the only series of reasonable length, the analysis here will concentrate on this particular category.

INVENTORIES AND PRICE DETERMINATION

For the purpose of this study, uranium consumption is defined as the amount of uranium entering the conversion–enrichment–fuel fabrication pipeline. Thus, consumption is largely determined by the current and short-term future level of installed nuclear-generating capacity. The rigid nature of U.S. enrichment contracts that were in force after 1973, combined with the relatively recent experience of reactor deferrals, delays, and cancellations, has meant that U.S. uranium consumption has been higher than warranted by the level of domestic nuclear-generating capacity alone. This has been reflected in a build-up of uranium inventories. Thus, actual consumption as defined here will differ from reactor consumption, the difference being accounted for by variations in the level of inventories of natural UF_6, enriched uranium, and fabricated fuel. Because U.S. consumption data are not generally available, total uranium deliveries (of both domestic and foreign origin) to U.S. domestic buyers was used as a surrogate variable in this study.

In the short term, the role of inventories is central to the process of price determination. Inventory holders have some idea of the ratio of inventories to consumption that they wish to maintain, usually referred to as the "desired" level of inventories relative to consumption. The effect on price operates through the relationship between the desired and the actual level of inventories. If the actual level of inventories is in excess of the desired level, then price will fall as attempts are made to unload the excess onto the market, and vice versa. Figure 8.4 illustrates the relationship between the short-term price of uranium (NUEXCO's exchange value) and the relative level of uranium inventories since 1971. The latter

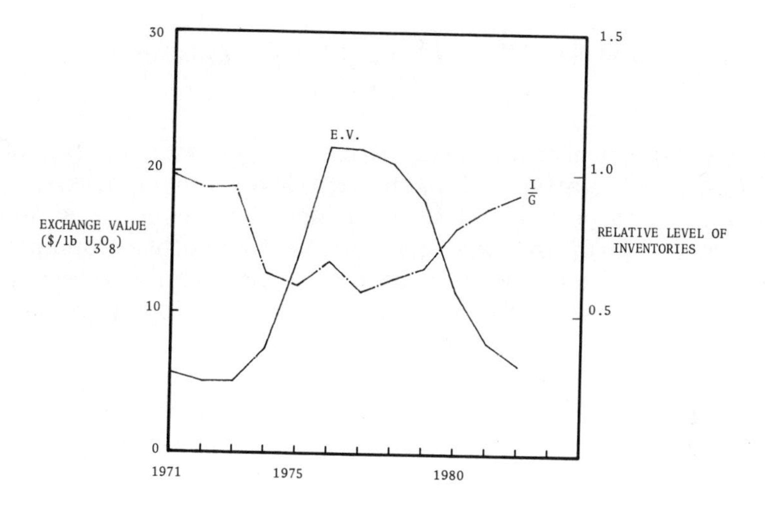

FIGURE 8.4 Relationship between the Exchange Value and the Relative Level of U.S. Uranium Inventories.
Source: Inventory data: *Survey of United States Marketing Activity,* U.S. Department of Energy, various issues. Exchange value: NUEXCO reports. The exchange value was deflated by the U.S. Producer Price Index for Industrial Commodities (1967 = 100).

variable is the ratio of total U.S. buyer inventories of natural uranium (from both domestic and foreign sources) to the level (in GWe) of U.S. nuclear capacity licensed to operate. The simple correlation coefficient between the average annual price of uranium and the relative level of uranium inventories at year-end of the previous year is –0.87, illustrating the high negative correlation between these two variables.

STATISTICAL ANALYSIS

Although the amount of data on the international uranium industry is very limited from the point of view of utilizing economic modeling techniques, it is possible to specify and estimate a simple relationship between the exchange value and the level of U.S. uranium inventories. The following equation was specified:

$$P_t = \beta_0 + \beta_1 P_{t-1} + \beta_2 \left(\frac{IS_t}{C_t} - \frac{IS_{t-1}}{C_{t-1}} \right) + u_t$$

i.e., short-term uranium prices are expressed as a function of the change in the relative (to consumption) level of inventories over the previous time period and its own lagged value. u_t is a random disturbance term. In practice, of course, many other economic and noneconomic factors would enter the relationship. In empirical work, however, it is often difficult, if not impossible, to isolate and measure the separate effects of all of them.

The above equation was estimated by the technique of least squares using data for the period 1971–82 and the following result was obtained:

$$\underset{(1.7768)}{P_t = 3.7102} + \underset{(6.9836)}{0.7449\, P_{t-1}} - \underset{(2.2355)}{8.4215} \left(\frac{IS_t}{C_t} - \frac{IS_{t-1}}{C_{t-1}} \right) + \underset{(4.2025)}{6.9261\, D_{75/76}}$$

$$\bar{R}^2 = 0.92 \qquad h = 1.12 \qquad SEE = 1.92$$

The estimated equation is reported with t ratios in parentheses under the corresponding coefficient estimates. Three summary statistics are provided: the coefficient of determination corrected for degrees of freedom, $\bar{R}^2$; the value of Durbin's h statistic for testing the hypothesis of nonautocorrelated (first order) disturbances in equations containing lagged dependent variables; and the standard error of the estimate.

All variables had their expected signs and (except for the constant) were well determined. The dummy variable, $D_{75/76}$, was assigned the value of unity in 1975 and 1976 and zero elsewhere. This variable is designed to reflect the impact on uranium prices of the "short" position in which Westinghouse found itself in 1975. While the decision to renege on its supply contracts was not announced until September 1975, its short position had been common knowledge within the nuclear-power industry for a considerable time before the announcement. Prices had been rising during 1975, therefore, in expectation of a supply shortfall caused, at least partly, by the Westinghouse predicament. The major impact of the September 1975 decision on the spot price of uranium occurred in 1976. The coefficient of the dummy variable implies that the Westinghouse incident caused the price of uranium to exceed its expected value by about \$6.95/lb U_3O_8.

By making different assumptions regarding the time path of the ratio of inventories to consumption, it is possible to obtain predictions of price behavior from our estimated equation. In the next chapter, however, a complete economic model of the U.S. uranium market is specified and estimated. By combining the equation explaining spot-price determination with the other equations in the model, predictions can be derived for all of the principal variables of interest in the U.S. uranium market for different assumptions regarding the time path of the ratio of inventories to consumption.

NOTES

1. A detailed history of the exchange value and changes in its definition, usage, and derivation is contained in *The Development of the Exchange Value Definition*, NUEXCO, November 1981.
2. The most recent report is *1982 Survey of United States Uranium Marketing Activity*, Energy Information Administration, Department of Energy, Washington, D.C., September 1983.
3. This section of the chapter is based upon the relevant section of the publication cited in note 2.

9

An Economic Model of the U.S. Uranium Market

INTRODUCTION

Since the inception of a commercial market for uranium in 1968, the United States has not only been the principal uranium producer in the world, it has also accounted for more than half of the total annual consumption and has thus been the dominant influence in the industry. Although this situation will change during the 1980s, with Canada becoming the world's principal uranium producer and with Western European nations (combined) consuming more than half of the world's annual requirements of uranium, nevertheless the United States will remain a significant force in the industry. In this chapter we develop a short-term economic model of the U.S. uranium market, using annual data for the period 1966–82. The model consists of five stochastic equations explaining uranium consumption, forward commitments, mine production, contract prices, and spot prices. An identity closes the model by allowing the level of uranium inventories to be determined. A short-term forecasting exercise is also undertaken.

Ideally, we would have liked to develop an economic model of the international uranium market, but much of the relevant information that would be required is unobtainable. For example, the role of inventories is central to the process of short-term price formation, yet few countries publish details of their uranium inventories. A further problem arises with the measurement of uranium consumption. Consumption statistics for individual countries, outside the United States, are kept confidential. Even for the United States, such data have only been made public since 1977. Paucity of non-U.S. data, therefore, necessarily restricts this model to the U.S. market.

A previous attempt at an economic model of the U.S. uranium market was made by Ahmed.[1] Using data for the period 1965–75, he specified and estimated separate equations explaining exploratory drilling, additions to reserves, uranium supply, and the price of uranium.

Of particular relevance to the current study is Ahmed's price equation. Ahmed specified both a uranium-demand and a uranium-supply equation in terms of the price of uranium and a number of other variables. By assuming the market to be in equilibrium (i.e., uranium demand and supply are equivalent), Ahmed was able to solve these two equations to yield a single equation with the price of uranium (*PU*) as the dependent variable. Upon estimation, Ahmed obtained the following result to explain uranium price formation in the U.S. market:

$$\log PU_t = 8.106 - \underset{(-2.79)}{0.5321} \log RR_t + \underset{(0.0370)}{0.00508} \log NCP_t + \underset{(0.163)}{0.03111} \log PC_t + \underset{(1.85)}{0.2735} \log WPO_t$$

$$R^2 = 0.76 \qquad DW = 1.30$$

where *RR* refers to total U.S. uranium reserves (irrespective of cost level), *NCP* to cumulative U.S. nuclear capacity (i.e., the sum of nuclear capacity on order, under construction, and operational), *PC* to the price of coal, and *WPO* to the world price of oil. The figures in parentheses are t ratios. Only the first of these four variables is significantly different from zero at a 5 percent level of significance, a fact which Ahmed blamed upon the presence of multicollinearity. Autocorrelated residuals also appear to represent a problem in this equation,[2] suggesting perhaps that the model is misspecified.

The equation presented above bears little resemblance to the price equation estimated in the previous chapter. Ahmed's price variable is a hybrid term, being comprised of government-controlled prices prior to 1968 and a weighted average of immediate and long-term private contract prices after 1970. Between 1968 and

1970 a combined price based on both government and private purchases was used. At no stage in his analysis did Ahmed consider the influence of uranium inventories on price formation. His equation therefore contains little explanation; it simply portrays uranium prices as monitoring changes in other fuel prices and the total level of U.S. uranium reserves. In fact, the nexus between uranium, oil, and coal price movements was broken during the latter years of the 1970s, as was illustrated quite clearly in Figure 3.2, and it is very doubtful whether Ahmed's price equation would achieve such a high level of explanation if his data period were extended beyond 1975.

THE U.S. URANIUM MARKET

The U.S. uranium market can be regarded as virtually a closed market over the period under study, since international trade in uranium played only a very minor role. Between 1966 and 1982, U.S. exports of domestic-source uranium amounted to only 23,400 tons U_3O_8. In the case of imports, the embargo on the enrichment of foreign uranium for domestic use was partly relaxed in 1977, although importation of yellowcake was permitted prior to that date. Between 1975 and 1982, a total of only 23,050 tons U_3O_8 was imported, 8,550 tons (37 percent) of which was imported in 1982 alone.

As U.S. domestic uranium requirements increase during the 1980s, however, domestic production is projected to lag significantly behind domestic requirements. As a consequence, the United States will place a greater reliance on imported uranium than at any time since the 1950s. With the ultimate lifting of the embargo on the enrichment of foreign uranium due to be completed in 1984, the degree of penetration of foreign supplies of uranium into the U.S. market should increase appreciably. Existing contracts and an excessive level of inventories, however, will mean that the bulk of this market penetration is unlikely to occur before the end of the 1980s. Table 9.1 details domestic deliveries since 1977 of uranium to DOE enrichment plants. It can be seen that foreign-source uranium accounted for only 18.1 percent of the total in 1982 even though its allowable level was 60 percent. To some extent, this position can be attributed to the slump in deliveries that occurred after 1979, following a relaxation of some of the more rigid conditions contained in U.S. enrichment contracts, but the high

TABLE 9.1 Deliveries of Uranium to DOE Enrichment Plants by Domestic Customers

	Origin (thousand tons U_3O_8)			Percent Foreign	
Year	U.S.	Foreign	Total	Actual	Allowable
1977	14.25	0.70	14.95	4.7	10
1978	11.95	0.75	12.70	5.9	15
1979	15.45	1.60	17.05	9.4	20
1980	11.15	1.20	12.35	9.7	30
1981	10.05	1.15	11.20	10.3	40
1982	13.55	3.00	16.55	18.1	60

Source: 1982 Survey of United States Uranium Marketing Activity, Energy Information Administration, Department of Energy, Washington, D.C., September 1983.

level of U.S. inventories that have accumulated since the late 1970s is undoubtedly the major explanation for lack of U.S. market penetration by the world's uranium-exporting nations.

Over the decade of the 1980s, U.S. exports of uranium will diminish to a negligible quantity while imports are projected to rise at a relatively steady rate. Since January 1, 1983, a total of 44,400 tons of foreign U_3O_8 has been committed for delivery to U.S. buyers through the year 2000. Over the same period, 127,750 tons has been committed for domestic delivery from domestic suppliers, while an estimated 48,250 tons is reported as unfilled requirements to the year 1992. This latter quantity will be met largely from inventory reduction and imports. There is also scope for U.S. producers to meet their domestic commitments by importing foreign uranium rather than mining it (possibly at a loss) themselves. If this were to become a common occurrence, one would expect a more rapid decline in U.S. production than is foreseen in Table 6.2.

Deliveries of uranium (from both domestic and foreign sources) to domestic buyers totaled 20,100 tons U_3O_8 (imports accounted for 8,550 tons) in 1982. Since deliveries are principally to meet enrichment feed contracts, they are closely related to the level of operational nuclear capacity, although after 1978 reactor cancellations, deferments, and licensing delays tended to reduce actual requirements substantially below the level of deliveries. As a consequence, inventories of both U_3O_8 and enriched uranium rose rapidly after 1978. Total U.S. inventories increased from 44,700 to 63,400 tons U_3O_8 equivalent between (year-end) 1977 and 1980. By the end of 1981, however, the situation had stabilized, largely due to a more flexible approach to enrichment contracts, which had allowed stocks of enriched uranium to be reduced. Nevertheless, the level of U.S. inventories at the beginning of 1983 (given in Table 8.4) was still excessively high relative to reactor requirements. The anticipated decline in the level of U.S. inventories (in absolute as well as relative terms) over the 1980s was reflected in NUEXCO estimates, which were given in Table 7.8. It is important to note, however, that these estimates are based upon the assumptions of a relatively rapid rate of decline in U.S. uranium production in the early 1980s (as projected in Table 6.2) and no imports or exports additional to those already under contract. These figures, therefore, do not represent a forecast of future inventory levels since they do not include import and export commitments that have yet to be made. Unless additional import commitments are offset by additional export commitments, lower domestic production, or cancellation of existing import commitments, these figures can be regarded as the minimum to which U.S. inventories are likely to fall over the decade of the eighties.

By historical standards, the 1983 level of U.S. uranium production was relatively low (10,600 tons U_3O_8), and with many producers facing uneconomic prices mine closings are relatively common. Production can therefore be expected to continue to fall during the early 1980s from its record peak of 21,850 tons reached in 1980 (1981 = 19,240 tons U_3O_8, 1982 = 13,430 tons U_3O_8). The U.S. import embargo allowed many high-cost U.S. producers the opportunity to survive in a marketplace where free competition with imports would probably have resulted in their rapid demise. Thus, the post-1980 plunge in production is partly due to the potential (and actual) flow of low-cost imports from Australia and Canada, in addition to the current industrial recession, high inventory levels, and the cancellation or deferment of many nuclear-power plants at various stages of their construction.

ECONOMIC CONSIDERATIONS

As many of the reactions to changed economic conditions in the uranium industry are spread over a considerable time period, the corresponding equations in the model have been formulated in terms of distributed lags. For example, a change in the spot price of uranium in the current time period will influence mine production over a large number of future periods. In the short run, variations in mine production in response to a change in the price of uranium can only be implemented by varying working hours and capacity utilization, and thus production is relatively price inelastic. In the longer term, however, existing mines can be expanded (or contracted) and, if warranted, new mines can be developed from known reserves (which can take up to 10 years).

Thus, at a given price P_t, we assume that U.S. uranium producers have a "desired" level of mine production MP_t^*, and that this relationship can be written (in the linear form) as

$$\text{(a)} \quad MP_t^* = \alpha + \beta P_t + u_t$$

where u_t is a random disturbance term.

Clearly, MP_t^* is unobservable. It is reasonable to assume, however, that in any time period the change in mine production adjusts partly to its desired level, i.e.

$$\text{(b)} \quad MP_t - MP_{t-1} = \mu(MP_t^* - MP_{t-1}) \qquad 0 < \mu < 1$$

where μ (a constant) is the speed of adjustment.

Substituting for MP_t^* in equation (b) and simplifying we obtain

$$\text{(c)} \quad MP_t = \eta_0 + \eta_1 MP_{t-1} + \eta_2 P_t + e_t$$

where $\eta_0 = \mu\alpha$, $\eta_1 = 1 - \mu$, $\eta_2 = \mu\beta$, and $e_t = \mu u_t$. If the disturbances in equation (a) are not serially correlated, then neither will be the disturbances in equation (c), and equation (c) can be consistently estimated by least squares.

In this study, a partial adjustment mechanism was used to explain mine production, consumption, forward commitments, and determination of the contract price. These four equations were estimated by least squares using data for the period 1966–82.[3] The equation explaining short-term uranium price formation was estimated in Chapter 8.

All estimated equations are reported with t ratios in parentheses under the corresponding coefficient estimates. Three summary statistics are provided: the coefficient of determination corrected for degrees of freedom ($\bar{R}^2$); the value of Durbin's h statistic for testing the hypothesis of nonautocorrelated (first-order) disturbances in equations containing lagged dependent variables, or the value of the Durbin-Watson statistic if no lagged dependent variable is present; and the standard error of the estimate (SEE).

SPECIFICATION AND ESTIMATION OF THE MODEL

Specification and estimation of the four stochastic equations are discussed on an individual basis below.

Mine Production

Mine production of uranium was defined as total U.S. uranium concentrate production from uranium ore plus concentrate obtained by solution mining, by heap leaching, or as a by-product of another activity. Uranium ore is by far the most important source of concentrate, accounting for about 80 percent of the total in 1982.

Mine production (i.e., uranium supply) will largely be determined by the current and expected future price of uranium, the level of reserves relative to production, and the cost of extracting those reserves. Clearly, this analysis can be extended to allow for the determination of additional reserves through exploration expenditure which, itself, will be largely determined by price expectations.

Cost data associated with the uranium-mining industry, however, were not available before 1979. In an attempt to take account of changing costs, a variable representing an index of miner's wages was constructed but it proved to have little explanatory power.

The estimated equation was

$$MP_t = \underset{(2.0127)}{3.2155} + \underset{(5.0724)}{0.5915}\ MP_{t-1} + \underset{(4.1408)}{0.2594}\ P^s_{t-1}$$

$$\bar{R}^2 = 0.8252 \qquad h = -0.8734 \qquad SEE = 1.3924$$

where MP denotes mine production (thousand tons U_3O_8) and P^s denotes the spot price (\$/lb U_3O_8).

The coefficients have their expected signs and, with the exception of the constant, were well determined at the 5 percent level of significance. The speed of adjustment is low, reflecting the long lead times involved in expanding or contracting uranium production, with only about 41 percent (i.e., 1.0 − 0.5915) of the gap between the desired and actual levels of mine production being closed each year. The short- and long-term elasticities of production with respect to price were 0.1967 and 0.4814, respectively.

Consumption/Demand

In Chapter 7, uranium consumption was defined as the amount of uranium entering the conversion–enrichment–fuel fabrication pipeline. Consumption, however, while easy to define is difficult to measure. Since the required data are not readily available for the years before 1977 (if indeed they exist), deliveries of U_3O_8 by domestic producers to the private domestic market plus imports destined for domestic use were used as a surrogate variable in this study. This definition of consumption, therefore, ignores changes in inventories held by utilities as U_3O_8, UF_6, and enriched uranium. While data on inventories are available from 1970 onward, we have noted already that the series was subject to major definitional changes in 1971 and 1979. Thus, we are unable to correct consumption for changes in inventories. Consumption, therefore, is defined as annual deliveries of U_3O_8, which is itself determined to a large extent by enrichment contracts placed in the past on the basis of projections of future levels of nuclear-generating capacity far exceeding today's projections. Hence the buildup of inventories.

As a consequence of the rigid nature of U.S. enrichment contracts, the level of uranium deliveries during the late 1970s/early 1980s was maintained at a level considerably higher than the desired level (i.e., current consumption plus changes in the desired level of inventories). For example, during 1980 U.S. deliveries averaged about 340 tons U_3O_8 for each GWe of operational nuclear capacity, whereas average consumption requirements were nearer 200 tons per GWe. Making use of the partial adjustment model once again, therefore, the estimated equation is

$$C_t = \underset{(7.1569)}{4.3525} + \underset{(4.2745)}{0.4428}\, C_{t-1} + \underset{(4.1792)}{0.1106}\, G_t + \underset{(3.9731)}{3.5669} D_{71}$$

$$\bar{R}^2 = 0.9737 \qquad h = -0.9709 \qquad SEE = 0.7782$$

where C denotes the level of uranium deliveries, both domestic and imported (thousand tons U_3O_8), G denotes the level of operational U.S. nuclear-generating capacity (GWe) at the beginning of the relevant year, and D_{71} is a dummy variable assigned the value unity in 1971 and zero elsewhere. All coefficients have the correct sign and are well determined at a 5 percent level of significance.

A price variable was included in the original specification, but it was found to be insignificant and hence was omitted from the final equation. This result was not surprising. As was discussed earlier, uranium to meet current consumption (i.e., enrichment contracts) must be purchased virtually irrespective of the prevailing price. In addition, the price of uranium is such a small component in the total cost of electricity generated by nuclear power that one would expect price fluctuations to be of limited importance in determining variations in consumption.

The speed of adjustment is fairly low, with only about 56 percent of the gap between desired and actual deliveries being closed each year. This undoubtedly reflects the rigid terms of post–1973 enrichment contracts, which required utilities to deliver uranium to the DOE for enrichment according to a fixed schedule, irrespective of changes in their consumption requirements. In 1971, for the first time, the U_3O_8 market was entirely private, and a large, unexplained increase in deliveries occurred. The coefficient of D_{71} implies that deliveries were approximately 3,600 tons higher than expected.

The coefficient of G_t implies that for every increase of 1 GWe in nuclear-generating capacity, deliveries of U_3O_8 will increase by about 110 tons. While this figure is very close to the annual reload fuel requirements for Light Water Reactors, it is substantially less than average consumption per GWe in recent years. Since this equation is based upon a partial adjustment mechanism, by making the appropriate substitutions it can be shown that an increase in nuclear-generating capacity of 1 GWe will increase the desired level of deliveries by about 198 tons U_3O_8. This figure is very close to annual U.S. consumption requirements for the early 1980s published by NUEXCO.[4]

Forward Commitments

Subject to constraints imposed by their enrichment-service schedules, utilities can be reasonably flexible concerning the timing of their contractual commitments for future deliveries of uranium. Since, in general, forward commitments represent contracts for intermediate and long-term delivery, one would expect the contract (rather than the spot) price to be a major influence on a utility's purchase decisions. In this study, however, the coefficient of the price variable had the a priori incorrect sign, although it was not significantly different from zero, and was consequently omitted from the final specification. This result can probably be attributed to the rigid nature of U.S. enrichment contracts, introduced in 1973, which required at least eight years' notice of separative work requirements and, as a consequence, forced utilities to enter into forward supply contracts to ensure an uninterrupted supply of enriched uranium. Price tended to be of relatively minor importance compared with security of fuel supplies.

The estimated equation was

$$FC_t = \underset{(2.85)}{21.26} + \underset{(6.2711)}{0.5922}\, FC_{t-1} + \underset{(2.8012)}{0.2513}\, TGC_t + \underset{(3.90)}{49.20}\, D_{76}$$

$$\bar{R}^2 = .9625 \qquad h = -0.7987 \qquad SEE = 10.274$$

where *FC* denotes forward commitments (thousand tons U_3O_8) and *TGC* denotes total generating capacity (GWe), i.e., the sum of U.S. nuclear-generating capacity on order, under construction, and operational.

All variables had the correct sign and were well determined. The speed of adjustment is low, with only about 40 percent of the gap between desired and actual commitments being closed each year. This reflects the above-mentioned rigidity of enrichment contracts that were in operation after 1973.

The dummy variable for 1976 (D_{76}) reflects the jump in forward commitments brought about by the Westinghouse announcement in September 1975 that it was unable to honor its supply commitment, which amounted to approximately 35,000 tons U_3O_8. Utilities that had entered into supply contracts with Westinghouse were therefore forced to enter new contracts for future delivery. The coefficient of D_{76} implies that an additional quantity of 49,200 tons was contracted for as a result of the particular circumstances operating in that year. Thus, the Westinghouse failure forced many utilities into long-term supply contracts to avoid the possibility of a perceived shortfall in uranium availability.

In the United States, there have been no domestic orders for nuclear-power reactors since 1979, while over the same period a large number of existing orders have been canceled. In fact, between year-end 1975 and year-end 1982 the variable *TGC* fell from 236.7 GWe to 136.0 GWe. This trend appears likely to continue during the mid-1980s but at a diminished pace. Other things remaining constant, the coefficient of *TGC* implies that forward commitments would decrease by about 251 tons U_3O_8 annually for every decrease of 1 GWe in *TGC*.

Contract Price

Uranium to meet future reactor requirements may be obtained under a contract for future delivery many years before it is actually consumed. Indeed, one effect of the Long-Term Fixed-Commitment Contract was to force utilities to enter into such long-term agreements to ensure continuity of supply of both uranium and enrichment services. In turn, this provided a stimulus to the uranium-mining industry and full order books for U.S. government-owned enrichment plants.

Clearly the cost of extracting uranium will be a prime component in the pricing policy of uranium producers. In the long term, revenue must cover at least long-term marginal costs (plus a user cost) to enable the producer to survive. Thus, the amount of uranium that is potentially available for extraction at various forward-cost levels should be an important variable in explaining the determination of contract prices.

We noted in Chapter 4 that WOCA and U.S. reserves are published according to a number of forward-cost categories. Historical and current estimates of U.S. uranium reserves at various forward costs are given in Table 9.2. As costs increase due to inflation, there will be a movement of reserves from lower-cost categories to higher-cost categories. Unless prices move in a corresponding manner, this will clearly raise the cutoff grade (i.e., the lowest grade of ore which it is economically viable to exploit) and consequently lower the level of uranium reserves that are economically viable at the current market price. The influence of inflation has led to the discontinuation of the $8, $10, and $15 forward-cost categories in the United States.

Using one of these cost categories to represent the level of U.S. reserves over the period of estimation would completely ignore the influence of inflation on the level of reserves. For example, the $10/lb U_3O_8 forward cost figure at January 1, 1977 was equivalent to a forward-cost figure of just $4.78/lb if 1967 were used as the base year. Obviously, the true $10/lb (1967 prices) level of reserves for 1977, therefore, would be considerably higher than the 250,000 tons given in Table 9.2.

To allow for the effects of inflation on U.S. uranium reserves, a simple extrapolative technique was used to adjust the level of reserves to a level corresponding to the forward-cost figure adjusted for inflation. Taking January 1, 1967 as the base date, the forward-cost category of $10/lb U_3O_8 was adjusted, using the Producer Price Index (PPI) for Metals and Metal Products,[5] to incorporate the influence of inflation. Reserves for each cost figure were calculated by using the following formula, which simply adds to the $10/lb U_3O_8 reserves for the relevant year the proportion of the difference between the $10 and $15 forward-cost categories accounted for by inflation:

$$R_t = R_t^{10} + \frac{\text{PPI}/100 - 1}{\$15 - \$10} (R_t^{15} - R_t^{10}) \qquad t = 1967, \ldots, 1982$$

R_t denotes reserves in year t, while R_t^{10} and R_t^{15} denote reserves in the \$10 and \$15/lb U_3O_8 categories, respectively, in year t. When the forward (inflated) cost exceeded \$15, the extrapolation was performed on the \$15 and \$30/lb cost categories, and after 1980 the \$30 and \$50/lb categories were used. The resulting "hybrid" reserve estimates are given in the final column in Table 9.2.

The contract price for uranium will also be influenced by the amount of new commitments for forward delivery that are made each year. The change in the level of forward commitments from one year to the next should reflect the current state of the market and its short-term prospects. For example, in the boom years of the

TABLE 9.2 Historical and Current Estimates of U.S. Uranium Reserves[1] (thousand tons U_3O_8)

	Cost Categories (per lb)						
Year[2]	\$8	\$10	\$15	\$30	\$50	\$100	Hybrid[3]
1965	151	175					
1966	145	195					
1967	141	200					200.0
1968	148	190	248				193.0
1969	161	200	265				211.1
1970	204	250	317				272.4
1971	246	300	391				334.6
1972	273	333	520				420.9
1973	273	337	520				457.0
1974	277	340	520	634			536.6
1975	200	315	420	600			462.7
1976	—	270	430	640			494.1
1977	—	250	410	680	840		516.2
1978	—	—	370	690	890		534.5
1979	—	—	290	690	920		581.5
1980	—	—	225	645	936	1,122	606.9
1981	—	—	112	470	787	1,034	470.6
1982	—	—	—	205	594	894	208.1

[1] The reserves in each category include all reserves in the lower cost categories.

[2] As at January 1.

[3] See text for method of calculation.

Source: Statistical Data of the Uranium Industry [GJO-100(81)], U.S. Department of Energy, Grand Junction, Colo., January 1, 1982.

early-1970s, when large numbers of new nuclear-powered plants were being ordered, the level of forward commitments increased substantially from year to year, whereas it stagnated in the later years of the decade as orders dried up and construction delays lengthened. Since 1979 the level of forward commitments has declined substantially, reflecting reactor cancellations, the poor prospects for new reactor orders, and high inventory levels.

Finally, since many contracts now contain market-price settlement clauses, one would expect the exchange value to exert a strong influence on contract price determination for this particular form of contract.

The estimated equation was

$$\underset{}{\Delta P_t^{c}} = \underset{(1.2034)}{0.2148} + \underset{(3.5333)}{0.1851}\ \Delta P_{t-1}^{s} + \underset{(1.6243)}{1.7609}\ \frac{\Delta FC_t}{FC_{t-1}} - \underset{(4.4347)}{0.0087}\ \Delta R_t$$

$\bar{R}^2 = 0.6586 \qquad DW = 1.2577 \qquad SEE = 0.6129$

The Durbin-Watson statistic (DW) indicates that significant autocorrelation of the residuals is present in this equation, with a probability of 0.0219 that such a value for DW can occur under the null hypothesis of nonautocorrelated disturbances.[6] The equation was reestimated, therefore, using a Cochrane-Orcutt type transformation suggested by Beach and Mackinnon.[7] The reestimated equation was

$$\Delta P_t^{c} = \underset{(0.6337)}{0.1488} + \underset{(3.0930)}{0.1732}\ \Delta P_{t-1}^{s} + \underset{(2.4137)}{2.2457}\ \frac{\Delta FC_t}{FC_{t-1}} - \underset{(3.8524)}{0.0078}\ \Delta R_t$$

$\bar{R}^2 = 0.6941 \qquad DW = 1.5968 \qquad SEE = 0.5801$

All coefficients had their expected signs and were well determined at the 5 percent level of significance (with the exception of the constant term). Originally, the specification was not in first difference form, but since the coefficient of the one-period lagged, contract-price variable was not significantly different from unity, the above specification was considered appropriate.

As might have been expected, the (relative) change in forward commitments is the variable that can generate major changes in the contract price. The influence of the spot price on the contract price is relatively small, a result that might have been expected by observing the relationship between the two series as shown in Figure 8.3. It is likely that the widespread adoption of a floor price in market-price contracts has, to some extent, reduced the influence that the spot price has had on the settled market price. Changes in the level of reserves also only influence the contract price on a relatively minor scale.

Closing the Model

An identity is required to close the model, namely:

$$\Delta IS_t = MP_t + M_t - C_t - X_t - \Delta IP_t$$

i.e., the change in buyer inventories is the total U.S. supply of uranium (mine production plus imports [M]) less domestic consumption, exports (X), and the change in the level of producer inventories (ΔIP). We have already noted, however, that a data series for consumption is not available and deliveries had to be used as a surrogate variable. Since deliveries are composed of both consumption and changes in buyer inventory levels, the separate influence of consumption cannot be identified for the purpose of a forecasting exercise. In addition, data on ΔIP are unreliable, although its significance is very minor.

Forecasting with the Model

Clearly, the accuracy of any forecast based upon the preceding equations will depend heavily on the projected values of the explanatory exogenous variables used therein, namely, G, R and TGC. Since ΔIS cannot be determined through the above identity, it is also necessary to treat this variable as being exogenously determined.[8] The projected values assigned to G, R, TGC and the relative level of inventories (IS/C), together with forecasts to 1990 for the endogenous variables of our estimated equations, are presented in Table 9.3 and Figures 9.1 to 9.6.

All data for 1982 are actual values. Estimates of G and TGC through to 1990 can be made with a reasonable degree of accuracy. The level of installed U.S. nuclear-generating capacity (G) is expected to increase by a fixed annual amount (5 GWe) to 1990, reaching a figure of 105 GWe in that year. NUEXCO has estimated that U.S. nuclear capacity will be 108.3 GWe by 1990.[9] In deriving

TABLE 9.3 Forecasting Experiments

Year	G	C	TGC	FC	R
1982*	65.4	20.1	136.0	155.5	208.1
1983	70.0	20.5	125.0	147.5	200.0
1984	75.0	21.2	120.0	140.0	180.0
1985	80.0	22.0	118.0	134.3	160.0
1986	85.0	22.9	116.0	130.4	160.0
1987	90.0	23.9	114.0	127.6	160.0
1988	95.0	24.9	112.0	125.5	160.0
1989	100.0	25.9	110.0	123.7	160.0
1990	105.0	26.9	110.0	122.2	160.0

Year	IS/C	P^S	P^C	MP	Year	IS/C	P^S	P^C	MP
Experiment I					*Experiment II*				
1982*	2.87	6.37	12.29	13.43	1982*	2.87	6.37	12.29	13.43
1983	3.16	6.01	12.11	12.81	1983	3.01	7.28	12.11	12.81
1984	3.47	5.58	12.24	12.35	1984	3.16	7.87	12.46	12.68
1985	3.82	4.92	12.38	11.97	1985	3.32	8.23	12.78	12.76
1986	3.82	7.38	12.35	11.57	1986	3.32	9.84	12.93	12.90
1987	3.44	12.41	12.88	11.97	1987	3.16	12.39	13.33	13.40
1988	3.09	15.90	13.86	13.51	1988	3.00	14.29	13.88	14.36
1989	2.78	18.16	14.58	15.33	1989	2.85	15.62	14.33	15.42
1990	2.51	19.51	15.09	16.99	1990	2.71	16.52	14.68	16.39
Experiment III					*Experiment IV*				
1982*	2.87	6.37	12.29	13.43	1982*	2.87	6.37	12.29	13.43
1983	3.00	7.36	12.11	12.81	1983	3.01	7.28	12.11	12.81
1984	3.00	9.19	12.47	12.70	1984	3.16	7.87	12.46	12.68
1985	3.00	10.56	13.00	13.11	1985	3.32	8.23	12.78	12.76
1986	3.00	11.58	13.32	13.71	1986	3.49	8.41	12.93	12.90
1987	3.00	12.34	13.60	14.33	1987	3.66	8.54	13.06	13.03
1988	3.00	12.90	13.84	14.89	1988	3.84	8.56	13.19	13.14
1989	3.00	13.32	14.05	15.37	1989	4.04	8.40	13.31	13.21
1990	3.00	13.63	14.24	15.76	1990	4.24	8.28	13.40	13.21

All price variables are expressed in constant (1967) dollars.
*Actual Values.

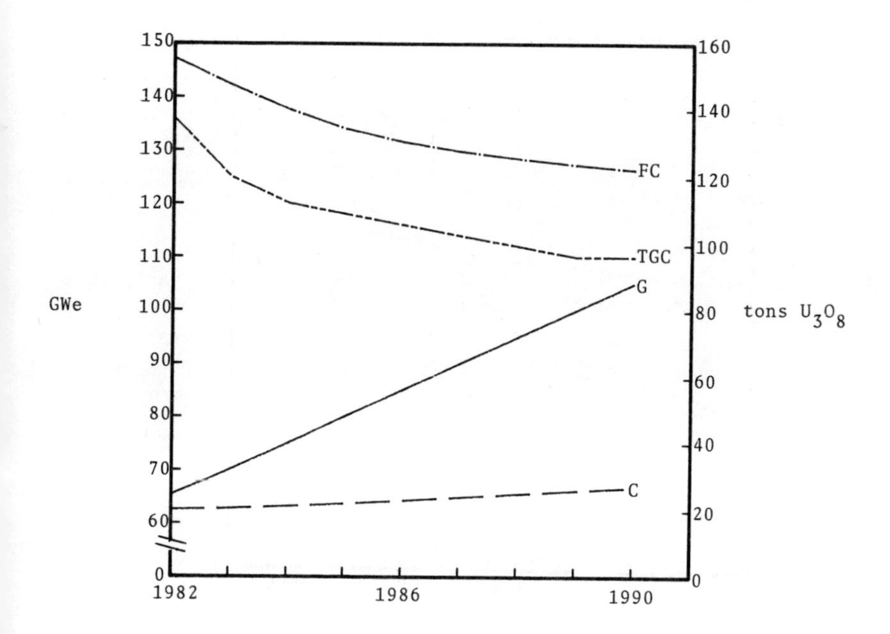

FIGURE 9.1 Projected Values for Deliveries (C), Forward Commitments (FC), Generating Capacity (G) and Total Generating Capacity (TGC).

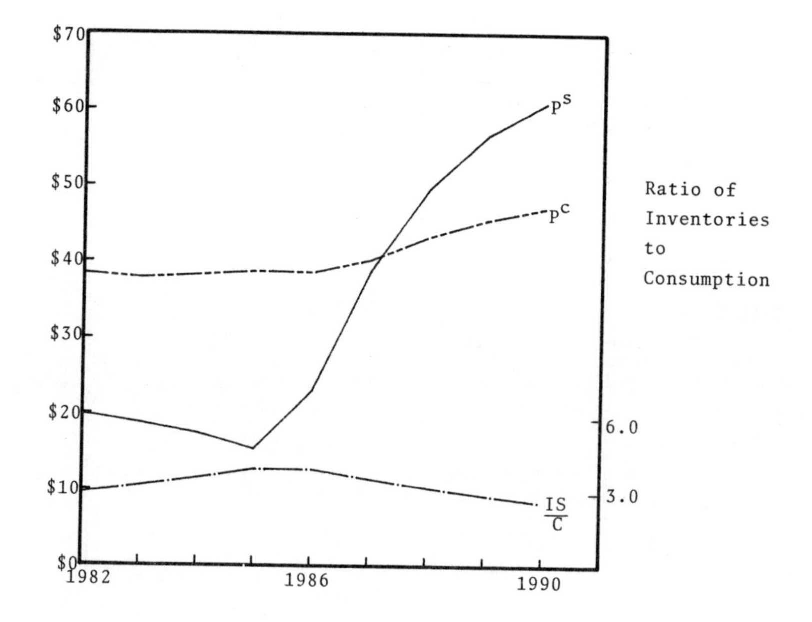

FIGURE 9.2 Projections for Experiment I.

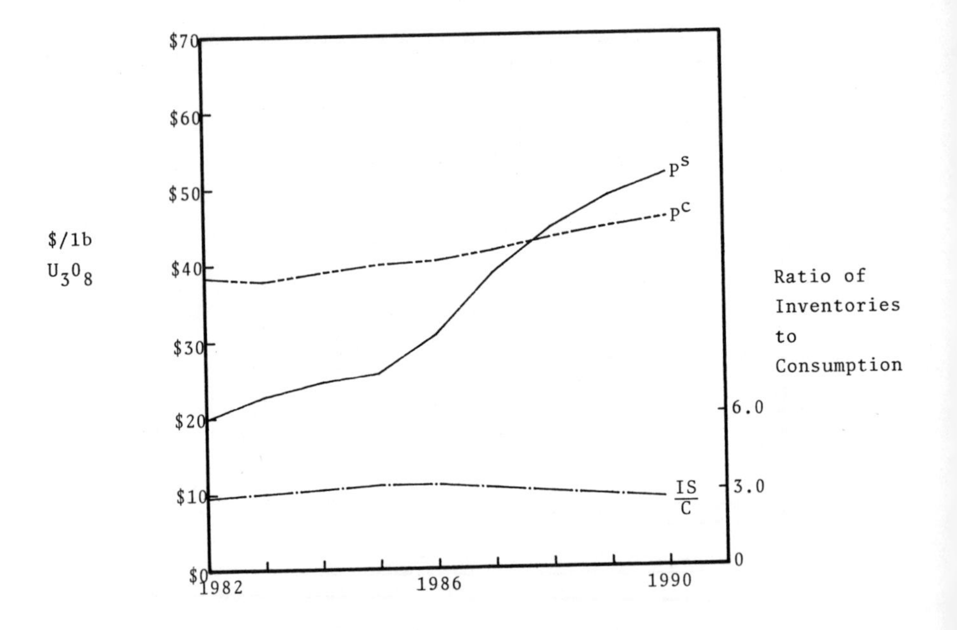

FIGURE 9.3 Projections for Experiment II (1982 prices).

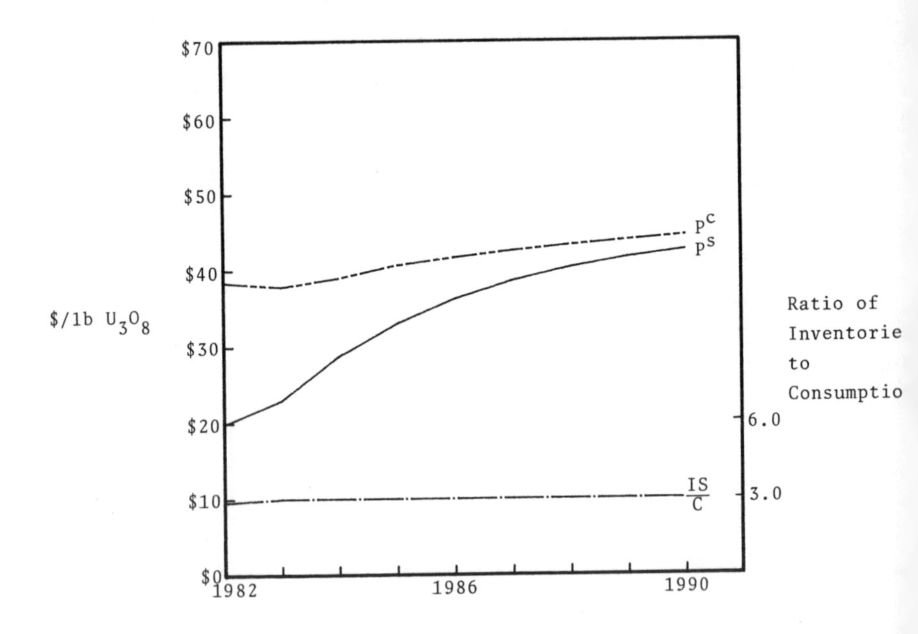

FIGURE 9.4 Projections for Experiment III.

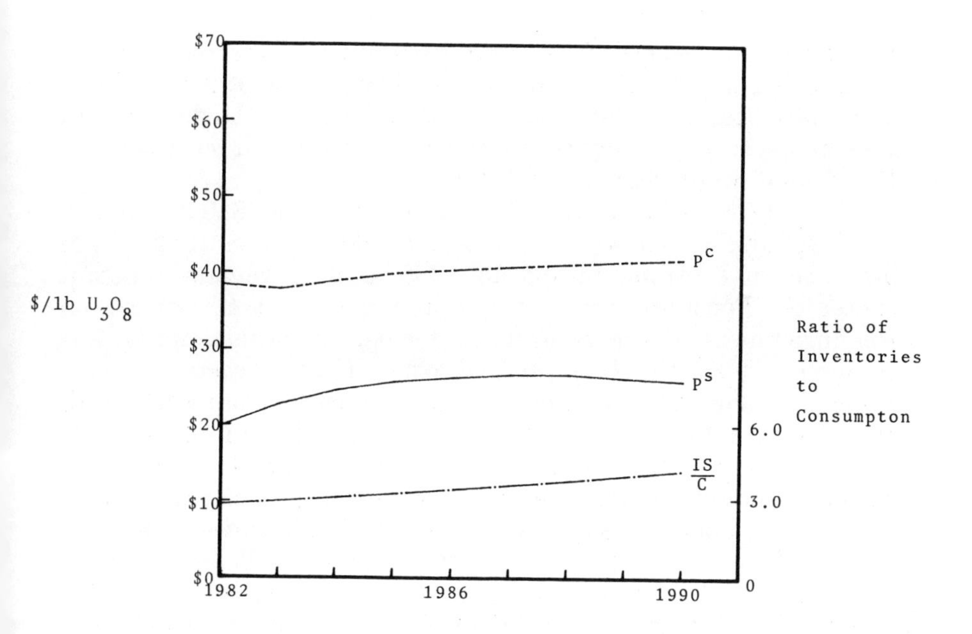

FIGURE 9.5 Projections for Experiment IV.

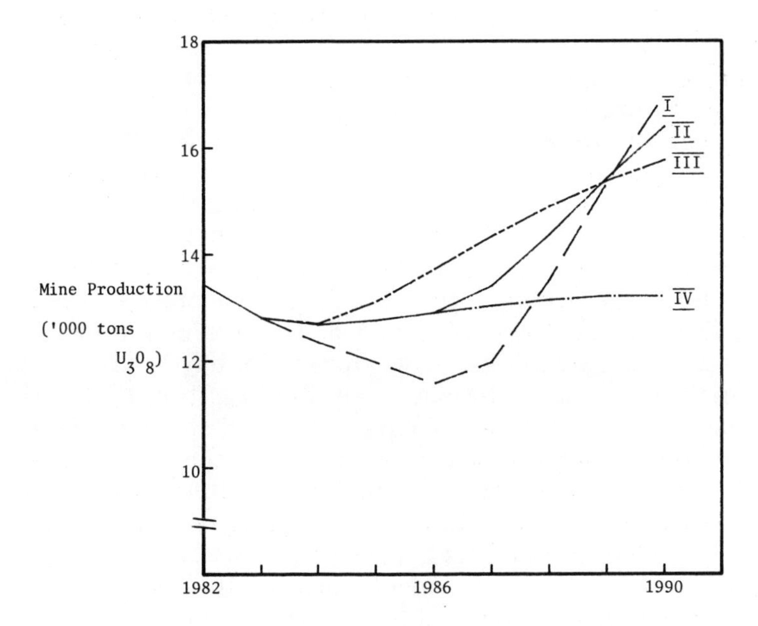

FIGURE 9.6 Projections for Mine Production.

projected values for *TGC* we have assumed that reactor cancellations will slow down over the early years of the 1980s and will eventually cease around the end of the decade. Under present conditions, it is unlikely that any new reactors will be ordered in the United States during the 1980s.

The level of low-cost U.S. uranium reserves (\$10/lb U_3O_8 in 1967 dollars, which was equivalent to \$30.16/lb in 1982), *R*, is assumed to continue falling to 1985, and to remain constant thereafter. For most of the estimation period, the U.S. embargo on the enrichment of foreign uranium for domestic consumption was in force. Thus, the level of low-cost domestic reserves was a principal variable in explaining contract-price determination. With the gradual removal of the embargo, imports of uranium began to feature prominently in supply transactions. By the mid-1980s, imports should account for a substantial proportion of new contracts and, as a consequence, the level of U.S. domestic reserves will be of little relevance for price determination. The assumption of a constant *R*, therefore, serves to remove the relevant variable (since $\Delta R = 0$) from the equation explaining contract prices.

As might have been expected, a rising value for *G* and a falling value for *TGC* generate corresponding movements in deliveries (*C*) and forward commitments (*FC*), respectively, as illustrated in Figure 9.1.

A major source of uncertainty in this forecasting exercise concerns the future time path of the relative level of inventories (*IS/C*). At year-end 1982, the ratio of buyer inventories of uranium to domestic deliveries stood at 2.87; i.e., buyer inventories were sufficient to meet nearly three years of deliveries at the 1982 level. Over the five-year period 1978–82, this ratio increased by an annual average of about 12.5 percent, mainly through a rapid and largely unintended accumulation of buyer inventories, which increased by 74 percent over this period. There is now evidence, however, of a slackening rate of growth in buyer inventory accumulation, with many market analysts suggesting that U.S. supply and demand will be in equilibrium in the mid-1980s. Thereafter, inventories will be drawn upon until they are reduced to a level more appropriate to consumption requirements. Two of the price projections given in Table 9.3, and illustrated in Figures 9.2 and 9.3, are based upon this hypothesis.

For each of the four experiments detailed in Table 9.3, and illustrated in Figures 9.2 to 9.5, different assumptions were made regarding the time path to 1990 of the relative level of inventories.

These time paths are illustrated at the bottom of each of the figures. Both price variables in Table 9.3 are expressed in 1967 dollars, whereas Figures 9.2 to 9.5 give price projections using 1982 dollars.

Experiment I. The inventories–consumption ratio (hereafter called "the ratio") is assumed to grow by 10 percent annually to year-end 1985, remain constant throughout 1986, and then decline by 10 percent annually to 1990. As a result of the rapid growth in inventories over the early years, the spot price is projected to fall sharply to 1985. Thereafter, however, a substantial recovery occurs as inventory levels fall rapidly. This probably represents the most extreme scenario that the U.S. industry could expect to face over the decade, with inventory levels fluctuating very sharply. Overall the spot price is predicted to rise by about 300 percent over the period 1982–90. Contract prices are also projected to rise over the forecast period but to a much smaller degree, about 23 percent. The latter, however, do not undergo the slump that spot prices experience up to 1986.

Experiment II. The ratio is assumed to follow the above pattern, but the rate of inventory accumulation and reduction is assumed to be 5 percent annually. This situation generates a modest increase in the spot price to the mid-1980s, leading to a fairly rapid rise thereafter as the level of inventories is reduced. The increase in the contract price is projected to be relatively moderate over the entire period.

Experiment III. The ratio is assumed to reach 3.0 in 1983 and to remain constant at that level over the remainder of the decade. This would imply that U.S. uranium supply and demand were in equilibrium and that, consequently, the actual and the desired levels of inventories were identical. If this were indeed the case, the spot price of uranium would ultimately reach a long-term equilibrium level independent of the level at which the inventories–consumption ratio remained constant. While such a result is of very limited importance, the equilibrium spot price for our equation is \$45.40/lb U_3O_8 (1982 dollars). It must be emphasized, however, that the above arguments and calculations are asymptotic and, in the context of the results obtained in this study, the long term is indeed very long. Short-term variations of demand and supply (and hence inventories) will be of far greater relevance for explaining changes in the spot price than will some distant asymptote. Again, contract prices are projected to experience a relatively moderate increase over the decade.

Experiment IV. The ratio is assumed to rise by a constant 5 percent annually, i.e., from 2.87 to 4.24 by year-end 1990. As a result, the spot price is projected to show a gradual rise to year-end 1988, before declining slightly to \$25.86/lb U_3O_8 (1982 dollars) by year-end 1990. The increase in the contract price is projected to be extremely small, but by the end of the forecast it is still considerably higher than the spot price.

It must be emphasized that the above possibilities are only for illustrative purposes. Clearly, the actual time path of the ratio of U.S. inventories to domestic deliveries will be more erratic than those shown in the above examples. In particular, the substantial price rises implied by Experiment I and II appear, with present knowledge, to represent rather optimistic projections. As a reference point, however, it is useful to note that our equation will yield a constant spot price through to 1990 if the ratio is assumed to grow at an annual rate of about 65 percent. Above this figure, the spot price would decline. It should also be noted that the projections are given in terms of constant dollars, and thus a forecast of the actual price that will prevail at some future date must also allow for the influence of inflation.

Figure 9.6 illustrates the response of U.S. mine production to changes in the spot price for the four experiments discussed above. After the mid-1980s, mine production is projected to rise as the result of increases in the spot price, but by historical standards all of the increases are relatively minor. These projections are, perhaps, the least satisfactory of the entire analysis as they disregard a major trend in the U.S. market. Unless an import embargo is reintroduced in the mid-1980s, it appears likely that U.S. production will stagnate in the face of competition from imports of relatively cheap Australian and Canadian uranium. The equation explaining mine production, and from which projections of mine production to 1990 were produced, was estimated, however, over a period when imports had a negligible impact on domestic production. Thus, the U.S. market will undergo substantial changes in supply patterns during the decade of the 1980s for which no allowance has been made in deriving the projections illustrated in Figure 9.6.

CONCLUSION

As it stands, the above analysis does not consider a number of important (nonquantifiable) factors that influenced the uranium market throughout the period under consideration in this study. The U.S. uranium market was subject to extensive government interference (in the form of the provision of enrichment services and an import embargo) throughout the period, as well as a wealth of political, social, and economic variables in addition to those already specified. Furthermore, the worldwide glut of uranium which became evident around 1980, as well as a surplus of conversion and enrichment service commitments, has led to the development of secondary markets in these three areas, thus introducing a degree of market flexibility that had previously been absent. Nevertheless, this analysis does serve as a basis for comprehending the importance and scale of the various economic variables which interact in the U.S. uranium market.

NOTES

1. Ahmed (1979).
2. With such a small sample size, almost inevitably the Durbin-Watson statistic will fall in the inconclusive region for the test. Ahmed did not provide an exact probability for the Durbin-Watson statistic in his equation.
3. All data used in this study, with the exception of the Producer Price Index, were taken from various issues of either *Survey of United States Uranium Marketing Activity* or *Statistical Data of the Uranium Industry* (both are publications of the Department of Energy). The Producer Price Index was obtained from *Monthly Labor Review*, U.S. Bureau of Labor Statistics.
4. *NUEXCO, Monthly Report on the Nuclear Fuel Market*, October 1983.
5. *Monthly Labor Review*, U.S. Bureau of Labor Statistics.
6. This probability was computed by the Pan Jie-Jian method (see Pan Jie-Jian, 1968).
7. Beach and Mackinnon (1978).
8. To treat ΔIS as being exogenously determined may not be too unrealistic. U.S. government enrichment and reactor licensing policies, to a large extent, caused the buildup in inventories in the late 1970s/early 1980s.
9. *NUEXCO, Monthly Report on the Nuclear Fuel Market*, October 1983.

10

The Economics of Uranium Enrichment

INTRODUCTION

On current projections, about 90 percent of all uranium required over the period to the year 2000 will have to be enriched. The guaranteed supply of enrichment services, therefore, is of critical importance to nations that rely on nuclear power as their major source of electricity. The provision of enrichment services, however, entails the construction of expensive plants using classified technology and, for plants using the diffusion process, considerable supplies of power. The enrichment stage of the nuclear fuel cycle accounts for nearly 50 percent of the total cost, excluding costs associated with the power plant itself, the other major cost being that of the raw material, yellowcake (U_3O_8). The costs associated with the other stages of the fuel cycle—namely conversion, fuel fabrication, and spent fuel storage, transport, and disposal—are all relatively minor compared with these two components.

In this chapter a number of factors that determine the demand for enrichment services are identified, and projections of enrichment capacity and requirements to the year 2000 are discussed.

ENRICHMENT SERVICES

Uranium-enrichment services are sold in separative work units (SWUs), which are a measure of the amount of effort required to separate U-235 from U-238. The proportion of U-235 remaining in the depleted uranium (the tails) after enrichment is called the tails assay.

In order to produce 1 kilogram of 3 percent U-235, 5.5 kilograms of natural uranium feed (i.e., 0.711 percent U-235) must be supplied to the enrichment plant, and 4.3 SWUs are utilized. The process also results in the production of 4.5 kilograms of 0.2 percent depleted U-235 (or tails). Schematically this process can be illustrated as follows:

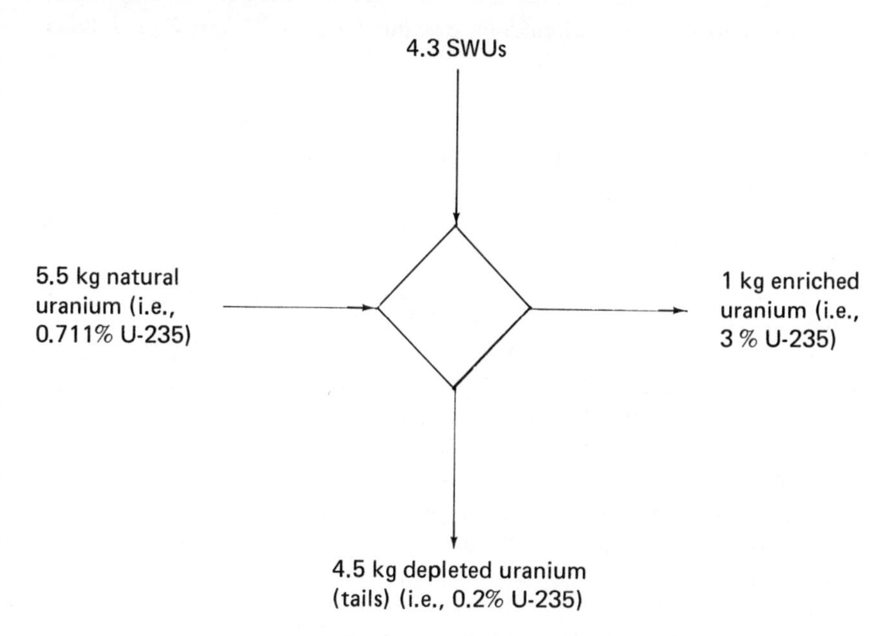

By decreasing the number of SWUs, it is possible to obtain the same amount of enriched uranium with a larger quantity of natural uranium (and vice versa). As a consequence, there would be a corresponding increase (decrease) in the tails assay since a lesser (greater) degree of "separation" will take place and hence the depleted uranium will contain an increased (reduced) percentage of U-235. Thus 6.6 kg of natural uranium feed combined with 3.4 SWUs would also have produced 1 kg of enriched uranium, but this time with 5.6 kg of depleted uranium with a 0.30 percent tails assay. Some examples of the effect of varying tails assay on the demand for natural uranium and SWUs are given in Table 10.1.

As the relative price of natural uranium to SWUs changes, a utility can minimize the cost of its enriched uranium requirements by adjusting its purchases of these two items accordingly. We have already noted that these two items combine to determine the tails

TABLE 10.1 Effect of Varying Tails Assay

Change in Tails Assay		Effect on the Demand for Natural Uranium and SWU Arising from Reactors Using Enriched Uranium
From:	*To:*	
0.20%	0.16%	- 6%
0.20%	0.25%	+ 9%
0.20%	0.30%	+20%

Source: The Uranium Equation, Uranium Institute, Mining Journal Books, London, April 1981.

assay. The tails assay corresponding to the minimum cost combination of uranium and SWU prices is called the optimum tails assay. Figure 10.1 illustrates the (nonlinear) relationship between the optimum tails assay and the ratio of the unit cost of feed (natural uranium) to the unit cost of SWUs.

It should be remembered, however, that current technology places a lower constraint of around 0.10 percent on the tails assay. If laser enrichment were to become a commercial reality this figure could be reduced to around 0.01 percent. In practice, there appears to be an upper limit of around 0.30 percent.

In the short term, a utility's flexibility with regard to achieving the optimum tails assay is limited by its contractual agreement with the enrichment plant which, in general, requires advanced notice of any customer change in tails assay requirements. In the longer term, however, it is apparent from the figures given in Table 10.1 that changes in the enrichment tails assay can have a very marked effect on the demand for both uranium and enrichment services.

Different enrichment agencies have different pricing schedules, and even within the one enrichment agency prices will vary according to the form of contract negotiated. Since uranium prices also vary according to contract conditions and the dates at which they were negotiated, it follows that there is no unique optimum tails assay. Rather, each utility will have its own optimum depending on the price it pays for uranium and SWUs.

During the early 1980s the spot price of uranium fell rapidly (see Table 3.2), while the cost of separative work undertaken by enrichment plants rose appreciably. The net combined result of these two price trends was that the optimum tails assay increased from around 0.20 percent U-235 in mid-1979 to more than 0.30 percent by mid-1983.[1]

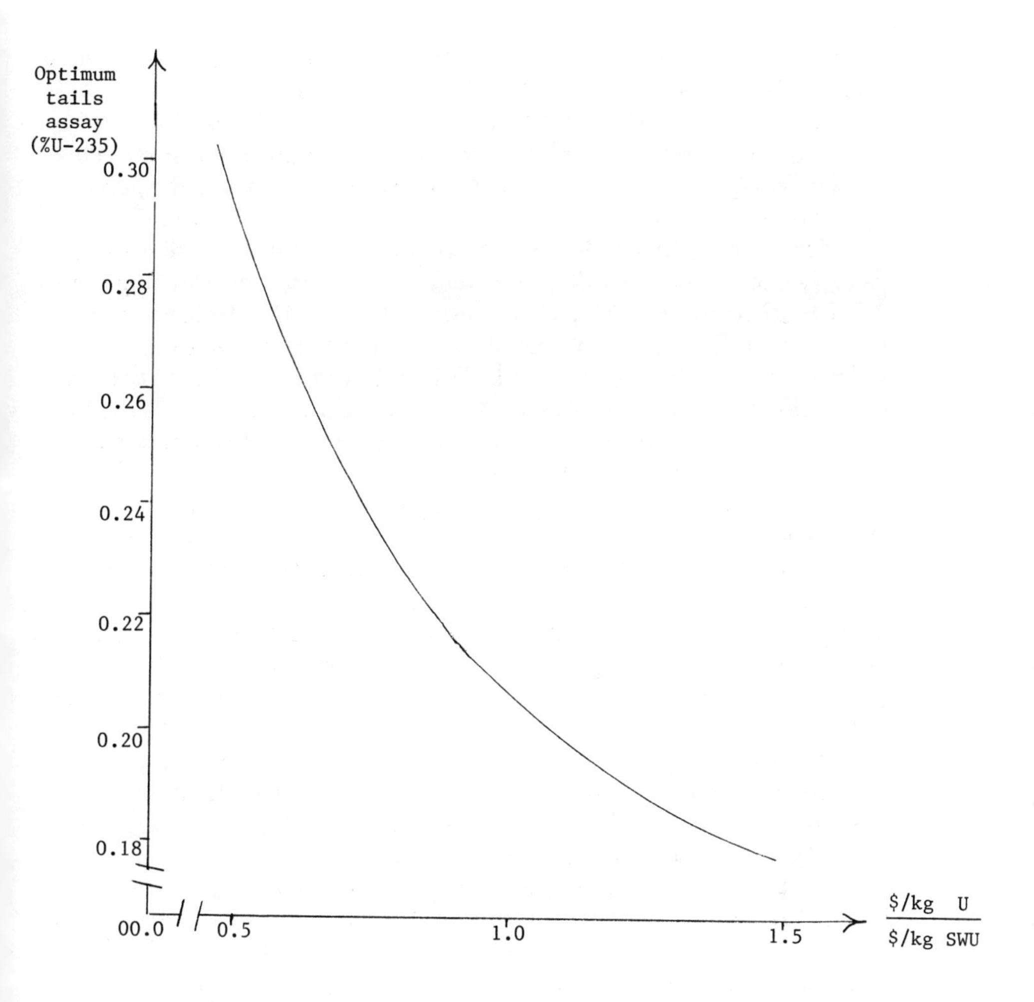

FIGURE 10.1 Optimum Tails Assay.

The optimum tails assay is a function of the ratio of the cost of a kilogram of natural uranium in the form of UF_6 to the cost of a kilogram unit of separative work.

For every 0.01 percent increase in the tails assay, natural uranium consumption increases by about 2 percent.

THE DEVELOPMENT OF AN ENRICHMENT INDUSTRY

In the aftermath of the Second World War, the United States and the USSR constructed gaseous-diffusion enrichment plants to satisfy their demand for highly enriched uranium for military purposes. While the Oak Ridge (Tennessee) enrichment plant was constructed during the war (1943), U.S. enrichment capacity was considerably expanded by the construction of additional plants at Paducah (Kentucky) and Portsmouth (Ohio) in 1951 and 1955, respectively. All three plants are government-owned but operated by private companies.

Until the mid-1970s the U.S. government, through the Atomic Energy Commission (AEC), the Energy Research and Development Administration (ERDA), and, more recently, the Department of Energy (DOE), had a monopoly on the provision of enrichment services to the Western world. This allowed it to encourage the expansion of nuclear power by providing U.S. utilities with all of their enriched uranium requirements at a very favorable price. Thus, any uncertainty attached to the availability of fuel was removed under the so-called "requirement" contracts that were in existence at this time.

Beginning in 1968, U.S. utilities were permitted to purchase U_3O_8 direct from the mines and toll its conversion and enrichment. In order to encourage utilities to enter into long-term contracts for enrichment services (and hence uranium supplies), in 1973 the AEC introduced the Long-Term Fixed-Commitment Contract (LTFC). By 1974 the AEC, having sold all forward production from its existing and committed enrichment facilities, closed its books, and this created a market for new suppliers of enrichment services.

Whether the LTFC was devised to ensure the long-term security of energy supply, or as a shot in the arm for the ailing U.S. uranium-mining industry, or as a gesture of encouragement to private (or non-U.S.) enterprise to enter the enrichment market is uncertain. It certainly initiated all three possibilities, although a somewhat less heavy-handed approach may, in retrospect, have been more conducive to the long-term stability of the uranium-mining industry.

During the years that ERDA enjoyed a monopoly on the provision of enrichment services, U.S. utilities were not free to select the tails assay they may have required. For many years ERDA pursued a "split-tails" policy in order to run down the large (50,000 tons U_3O_8) government stockpile. Thus the contractual tails assay of 0.20 percent (for which the utilities would deliver the necessary uranium feed and pay for the corresponding number of SWUs) was generally lower than the operating tails assay (i.e., that actually used by ERDA). The extra uranium that was required to operate this scheme came from the stockpile. As a consequence, utilities were paying for more SWUs than were actually used.

The sharp fall in orders for nuclear power reactors in the late 1970s, together with widespread cancellations and deferments of existing orders (especially in the United States), meant that U.S. enrichment plants no longer had full order books. The entry of two consortia of Western European nations (Eurodif and Urenco) into the market at this time, together with the willingness of the USSR to supply enrichment services to Western Europe, forced ERDA (now the Department of Energy) to alter its terms in the face of more flexible terms offered by these new competitors. The current status of the four major contributors to WOCA's enrichment capacity is given in Table 10.2.

European enrichment plants have adopted a commercial-pricing policy. The DOE has a cost-recovery selling-price policy. But cost (as defined by the DOE) does not include any significant taxes or appreciable capital amortization. Consequently, U.S. enrichment services have, in the past, been cheaper than the European product.[2] Depending on the type of contract, at September 1983 the DOE price for enrichment was in the range of \$138.65–\$149.85/SWU, whereas Eurodif and Urenco services were sold at \$170–\$180/SWU. On the secondary market, however, NUEXCO[3] estimated the value of enrichment services for 1984 to be \$90/SWU, a figure which reflects the saturated condition of the enrichment market.

Eurodif, a joint venture between France (the major partner), Belgium, Italy, and Spain (Iran was originally a member), has a gaseous-diffusion plant at Tricastan in France. This plant attained full design capacity in 1982 with member nations committed to taking delivery of enriched uranium in proportion to their shares in the project. Tricastan's design capacity, however, is sufficient to supply all of Western Europe's enrichment requirements during the mid-1980s, and consequently the plant is currently operating considerably below full capacity.

More recently Urenco, a joint U.K., Dutch, and FRG venture, was established following the signing of the Almelo Treaty in 1970. Small pilot plants using centrifuge technology were then in operation. Two demonstration plants were commissioned in 1976, and larger commercial units are now in operation. Since centrifuge technology allows the gradual expansion of enrichment plants, future expansion plans will depend upon market conditions.

Expansion of U.S. enrichment capacity at Portsmouth is planned for the late 1980s. The first two increments of 1.1 million SWU capacity of an add-on gaseous centrifuge plant are scheduled for completion in 1988 and 1989. Additional increments of 1.1

TABLE 10.2 Current (1982) Enrichment Capacity Available to the Western World and Tails Assay Flexibility

Agency	Location	Capacity (SWU/year)	Variable Tails Range	Notice of Alteration	Estimated or Reference Tails
DOE[1]	U.S.A.	27.3 million[2]	0.16–0.30%	15 months prior to delivery	0.20%
Eurodif	France	10.8 million	0.18–0.32%	4 years prior to year of delivery	0.25%
Urenco	Holland U.K.	600,000 400,000	0.20–0.30%	4 years prior to initial delivery	As set by customer
Techsnab-export	USSR	2–4 million[3]	0.20% upward	9 months prior to year of delivery	0.20%
	total capacity	41.1–43.1 million			

[1] These figures relate to the U.S. Department of Energy's adjustable fixed commitment contract. The requirements and long-term fixed-commitment contracts do not allow for a variable tails assay nor for notice of alteration. The draft of a new enrichment contract which has been designed to improve the competitive position of the U.S. Department of Energy has recently been released. It was due to be brought into practice on July 1, 1983, but has been delayed until early 1984.

[2] Nameplate capacity. Actual capacity available for commercial enrichment is only 25.4 million SWU/year.

[3] The figure given for Techsnabexport represents contracts with Western European utilities until 1990, *not* capacity. This capacity is unlikely to be available after 1990 due to increasing demands of Eastern European nations.

Source: Adapted from *The Uranium Equation*, The Uranium Institute, Mining Journal Books, London, 1981.

million SWU/year up to the nominal full capacity of 13.2 million SWU/year will be added as required by the enrichment market. Future expansion at Portsmouth will also depend on any decision to retire some of the less efficient U.S. gaseous diffusion facilities and on the progress with advanced enrichment technologies.

A second European gaseous-centrifuge enrichment agency (COREDIF) was planned by the Eurodif consortium. The proposed ultimate capacity was to have been 10 million SWU/year, but the project has been shelved due to an oversupply of enrichment capacity, which is projected to last well into the 1990s.

DETERMINANTS OF ENRICHMENT REQUIREMENTS

The fuel requirements of the current generation of Light Water Reactors (LWRs), assuming a two-thirds Pressurized Water Reactor (PWR) and one-third Boiling Water Reactor (BWR) mix, necessitate an average separative work requirement of a little over 100,000 SWU/GWe per annum (assuming a 70 percent load factor and a 0.20 percent tails assay). Given projections of installed nuclear capacity to the year 2000, it is therefore possible to forecast the corresponding level of separative work requirements for a range of alternative assumptions regarding:

1. Reactor mix,
2. Recycling and reprocessing,
3. Tails assay,
4. Stockpile policy,
5. Political factors.

These five factors will now be considered individually.

Reactor Mix

OECD[4] projections of installed nuclear capacity to the year 2025 indicate that LWRs, and in particular the PWR, will increase their already substantial share (currently about 87 percent) of the WOCA market using a once-through fuel cycle (i.e., no recycling of spent fuel). Improved technology with the PWR program could lead to a reduction in both uranium and SWU requirements per GWe, but the effect on the latter is unlikely to be very noticeable before the end of this century.

Over the same period, the significance of reactors that do not make demands on enrichment capacity is projected to remain very minor, with only the Pressurized Heavy Water Reactor (PHWR), the Fast Breeder Reactor (FBR), and the Gas-cooled Reactor (GCR) being operational. Only Canada will have a significant capacity in PHWRs, while the GCR in the United Kingdom—and its technological successor, the Advanced Gas-cooled Reactor (AGR), which requires its fuel to be enriched—will gradually be superceded by a PWR program.

By the year 2000, only France is projected to have a significant FBR capacity operational. Widespread adoption of the FBR, however, would eventually reduce the demand for SWUs, but this appears an unlikely event until well past the turn of this century.

Recycling and Reprocessing

Spent fuel can be reprocessed to separate the residual uranium and reactor-produced plutonium from the waste products generated as a result of fission in the reactor. The recovered uranium reenters the fuel cycle at the conversion stage, and thus its recovery directly reduces the demand for uranium. In addition, enrichment requirements (SWU) would be reduced as recovered uranium is still slightly enriched (around 1 percent U-235). Recovered plutonium, however, reenters the cycle at the fuel-fabrication stage either to produce fuel for FBRs or, in the future, for use in plutonium-burning or mixed oxide LWRs. Thus plutonium recovery directly influences the demand for both uranium and enrichment services, but it is unlikely to be a factor of any consequence until well past the year 2000.

Tails Assay

We have already noted the considerable impact that can be made on the demand for both uranium and enrichment services by varying the enrichment tails assay. The OECD assumption of a 0.20 percent tails assay is, at present, substantially below the optimum tails assay (based on the cost of U.S. enrichment services and the U.S. spot price for uranium). This situation, induced by a worldwide glut of uranium, is likely to continue through at least the mid-1980s. At year-end 1983, the optimum tails assay was slightly higher than the maximum limit (0.30 percent) that is permitted by most enrichment agencies.

Stockpiling Policy

While different utilities and different nations will have varying strategies regarding the optimum level of stockpiles, stocks amounting to approximately 2–3 years of forward requirements are generally regarded as ideal. Currently WOCA stocks amount to approximately 4–6 years of WOCA forward consumption (but are slowly declining), which is excessively high by any yardstick.

Political Factors

U.S. nonproliferation policy has, in the past, attempted to prevent the spread of enrichment technology to politically sensitive areas. Thus the supply of enrichment services was driven more by political and strategic factors rather than economic ones. Recently, however, Brazil has acquired enrichment technology from the Federal Republic of Germany, while South Africa has developed its own enrichment process. U.S. nonproliferation policy with respect to enrichment technology appears, therefore, to have been a dismal failure.

PROJECTIONS OF ENRICHMENT REQUIREMENTS

On the basis of the most likely combination of the above factors, the OECD has produced projections of future levels of enrichment requirements based upon projections of the rate of growth of nuclear power in WOCA nations.

The OECD's projections of nuclear-power growth to the year 2025 for the four major regions of WOCA are summarized in Table 10.3. While the *maximum* estimates for the years to 1990 are relatively fixed due to the long lead times involved in the planning, construction, and licensing of nuclear-power reactors, the lower boundary may be subject to significant variation as projects are expanded or contracted. Currently it appears that even the low estimates for 1985 and 1990 are overly optimistic.

Beyond 1990 the degree of uncertainty associated with the projections in Table 10.3 is reflected in the substantial difference between the low and high estimates. Such a degree of uncertainty is warranted given the dramatic revisions that have occurred in projections made over the past decade of nuclear-power growth. This point was illustrated in Figure 7.1, which showed the plunge that took place during the 1970s in the anticipated rate of growth of nuclear power. This variability must be borne in mind when considering the current projections.

Table 10.4 provides estimates of separative work requirements corresponding to the nuclear-capacity projections given in Table 10.3. The data for OECD America refer exclusively to the United States, since the current generation of Canadian PHWRs does not require its fuel to be enriched. The OECD nations in Europe that will require substantial amounts of separative work

TABLE 10.3 Projections of WOCA Nuclear Power Growth (GWe) (installed nuclear capacity, year-end)

	1980	1985	1990	1995	2000	2025
OECD Europe	47	94–95	142–158	171–219	223–317	457–911
OECD America	58–60	96–119	138–156	157–185	185–235	289–643
OECD Pacific	15	28–30	51–53	67–84	89–131	169–360
Developing WOCA	3	14	30–33	56–72	88–121	396–880
WOCA	123–125	232–258	361–399	451–560	585–804	1,311–2,794

Source: Adapted from OECD Nuclear Energy Agency, *Nuclear Energy and Its Fuel Cycle: Prospects to 2025*, OECD, Paris, 1982.

TABLE 10.4 Projections of Annual Separative Work Requirements (million SWU/annum) (0.20 percent tails assay, 70 percent load factor)

	1980	1985	1990	1995	2000
OECD Europe	8	11–13	14–16	19–25	24–36
OECD America	7	11–12	15–16	15–18	18–24
OECD Pacific	2	4–4	6–7	8–11	10–16
Developing WOCA	1	2–2	4–5	7–9	11–16
	18	28–31	39–44	49–63	63–92

Source: As for Table 10.3.

during the remainder of this century are Belgium, the Federal Republic of Germany, France, Italy, Spain, Sweden, Switzerland, and the United Kingdom. France will dominate this group, however, and should account for about 50 percent of Western European requirements. OECD Pacific data refer exclusively to Japan, while the developing WOCA data are dominated by Southeast Asian nations.

Overall, SWU requirements by the year 2000 are projected to be at least 350 percent higher than 1980 requirements, with the major (potential) growth occurring in the decade of the 1990s.

PROJECTIONS OF ENRICHMENT CAPACITY

Projections of uranium-enrichment capacity to the year 1995 are given in Table 10.5. This table does not take account of the small enrichment plants currently under construction in Brazil and South Africa, nor does it consider the possible or planned entry into the enrichment market of Australia and Canada.

By comparing Tables 10.4 and 10.5, it is apparent that current enrichment capacity is far in excess of current requirements and is likely to remain so until the turn of the century if all expansion plans come to fruition. If Urenco continues with its planned expansion, then by 1990 all major Western European uranium-consuming nations (with the exception of Sweden and Switzerland) will be self-sufficient in enrichment requirements. Current U.S.

TABLE 10.5 Projections of Uranium Enrichment Capacity (million SWU/year)

	1980	1985	1990	1995
France[1]	6.0	10.8	11.8	15.8-16.8
Germany (FRG)[2]	—	0.4	1.0	—
Holland[2]	0.2	1.1	3.6	6.0-6.9
United Kingdom[2]	0.5	0.8	3.3	5.5-6.6
United States	26.4	27.3	30.6	36.1
Japan	0.02	0.35	2.5	5.5
total	33.12	40.75	52.8	74.9-78.4

[1] Eurodif plant at Tricastan (France)
[2] Joint partners in Urenco

Source: As for Table 10.3.

enrichment capacity greatly exceeds its domestic requirements, and this situation is expected to continue throughout the 1980s. Clearly the demand for enrichment services from nations of the developing world is an important consideration in U.S. expansion plans, but the Western European enrichment agencies will also be in a position to meet this demand.

While Japan has plans to expand its domestic enrichment industry beyond the gaseous-centrifuge demonstration plant at Ningyo-Toge, it will remain a substantial net importer of enrichment services over the remainder of this century.

The total figure for 1985 given in Table 10.5 represents firmly committed capacity (i.e., either operational, under construction, or on order). Projections for 1990 and 1995 involve greater uncertainty as there is no guarantee that this additional capacity will be built. The 1985 figure, therefore, also represents minimum committed capacity for 1990 and 1995. Actual capacity may, in fact, fall below this figure if the U.S. Department of Energy decides to retire some of its high-cost gaseous-diffusion facilities. Overall, the post-1985 estimates are very uncertain and it is likely that most expansion plans will be shelved or canceled through lack of demand.

ASSESSMENT OF THE ENRICHMENT MARKET

The current level of WOCA uranium enrichment capacity is sufficient to satisfy the enrichment requirements (based upon OECD projections of installed nuclear capacity) of the WOCA nations until at least the mid-1990s. If expansion plans for the 1980s come to fruition, these will only serve to exacerbate this surplus supply situation.

OECD projections of separative work requirements are based upon projections of installed nuclear capacity, combined with assumptions regarding reactor mix, recycling and reprocessing policies, the level of the enrichment tails assay, stockpiling policy, and political factors. Of critical importance are basic assumptions concerning the optimal level of the enrichment tails assay and the load factor for installed capacity. It is now apparent that not only are the recent OECD projections of installed nuclear capacity to the year 2000 overly optimistic, but that the level of separative work requirements has also been overestimated because of errors in these two basic assumptions.

The OECD's low projections for installed nuclear capacity through to the year 2000 given in Table 7.2 appear to represent overly optimistic forecasts for several major uranium-consuming nations. In particular, it is extremely unlikely that (given current lead times) U.S. installed nuclear capacity will exceed 150 GWe by the year 2000, and similarly Japan is unlikely to exceed a level of 75 GWe by the same year. WOCA installed nuclear capacity by the year 2000, therefore, is likely to be substantially below the projections given in Table 7.2, with a corresponding reduction in enrichment requirements.

A combination of falling uranium prices and the rising cost of separative work forced the optimal tails assay up to around 0.3 percent during the early 1980s. OECD estimates of future enrichment requirements, however, assume a 0.2 percent tails assay. If the average tails assay requested at the enrichment stage rises to 0.3 percent, this would represent a decrease of 20 percent in SWU requirements over those projected by the OECD. Thus, relatively low (by historical standards) uranium prices are encouraging a conservation of enrichment capacity, and this appears unlikely to change during the 1980s.

OECD projections of enrichment requirements also assume a 70 percent load factor which, on the basis of past experience, is too high. During the year ended December 1983 the load factor (weighted by reactor size) for WOCA countries was approximately 60 percent for both PWRs and BWRs, and this figure is reasonably representative for overall LWR performance over the past decade. Only the PHWR has consistently maintained a capacity factor close to 70 percent, but this type of reactor accounted for only about 4.5 percent of total WOCA installed nuclear-generating capacity in 1983 and does not require its uranium to be enriched.

The impact of this lower-than-anticipated load factor continuing through the 1980s would be to lower the demand for enriched uranium, and hence enrichment services, by approximately 14 percent.

A 60 percent load factor combined with a 0.30 percent tails assay would result in a fall of approximately 31 percent in enrichment requirements below the projected levels given in Table 10.4. This probably represents, however, a rather pessimistic estimate of future requirements. Following the accident at Three Mile Island, the U.S. load factor has been around 55 percent. Since the United States accounts for a substantial proportion of WOCA's installed nuclear capacity, its low load factor has served to lower the WOCA average significantly. From Table 7.5 it is also apparent that both France and the United Kingdom are experiencing low load factors. It is likely, however, that load factors in these three countries will improve significantly as various operational problems and retrofitting delays are overcome and, in the case of the United Kingdom, more efficient types of reactors (AGRs and PWRs replacing the relatively old Magnox GCRs) are commissioned. Nevertheless, when reductions in the level of projected installed nuclear capacity are also taken into consideration, it is apparent that OECD projections of the demand for enrichment services will still represent a substantial overestimate. It appears likely, therefore, that the committed level of enrichment capacity for 1985 (given in Table 10.2) will be sufficient to satisfy annual demand until at least the end of the century. Any expansion by existing or new suppliers of enrichment services over the next decade is likely to be largely at the expense of U.S. plants, some of which are nearing the end of their economic life. Thereafter, the possible introduction of laser enrichment as a commercially feasible technology may render redundant the current modes of uranium enrichment.

NOTES

1. About 10 to 15 percent of uranium requirements in the United States are traded on spot or short term. At the end of December 1982 the spot price was \$20.15/lb U_3O_8. The average contract price for deliveries in 1982, however, was \$38/lb. Thus the optimum tails assay for the bulk of U.S. consumers would be considerably below that of those obtaining their uranium supplies on the spot market.
2. The volatility of currency exchange rates during the early 1980s, and in particular the rapid appreciation of the U.S. dollar against the majority of European currencies, removed the price advantage that U.S. enrichment agencies had over their European counterparts.
3. *NUEXCO, Monthly Report on the Nuclear Fuel Market*, October 1983.
4. All references in this chapter to data published by the Organisation for Economic Cooperation and Development (OECD) refer to OECD Nuclear Energy Agency *Nuclear Energy and Its Fuel Cycle: Prospects to 2025*, OECD, Paris, 1982.

11

Prospects

INTRODUCTION

This book has been concerned with the economics of the world uranium industry, ranging from the birth of a private market in the United States in the late 1960s, through the "boom" years of the second half of the 1970s, to the "bust" years of the early 1980s. The industry's destiny, in common with its history over the last two decades, is, for all practical purposes, completely determined by the rate of expansion of the world's nuclear-generating capacity. Given the long lead times that are inherent in nuclear plant construction programs, the maximum level of uranium requirements through the early 1990s can be derived with a reasonable degree of certainty. Thereafter, however, the nuclear-power (and hence the uranium) industry's prospects will themselves be decided by the changing pattern of the world's energy requirements.

Through the early decades of the next century, world energy demand will be dependent upon a wealth of interacting economic, demographic, technical, and political factors. Given the great degree of uncertainty inherent in any forecasts of the future behavior of these variables, it follows that predicting long-term uranium requirements is likely to be a very difficult proposition. A number of events during the 1970s (the OPEC oil embargo, major oil price rises in 1973 and 1979, the Westinghouse case, the accident at Three Mile Island) combined to decimate growth estimates that had been made around 1970. Hindsight should act as a salient reminder of the pitfalls which could plague similar predictions of uranium requirements for the remaining years of the twentieth century and beyond.

PROJECTIONS TO 2025

Demand

The OECD/IAEA has produced estimates of world levels of energy demand to the year 2025, the share of nuclear power in that demand, and the resulting demands at each stage of the nuclear fuel cycle.[1]

On the basis of published statistics and questionnaire returns from member nations, electricity demand was projected to the year 2025. Nuclear's share of this demand was then estimated. Nuclear power was projected to account for (low–high range) 35–50 percent of electricity generation in OECD America, and 50–70 percent elsewhere. The resulting estimates for the long-term growth of nuclear power are given in Table 11.1.

As might be expected, the longer the time frame for the projections, the wider the gap becomes between the low and high estimates (i.e., a greater degree of uncertainty is present). Given that the likely figure of installed nuclear capacity in 1990 will probably be around 280 GWe, even the low estimate for the year 2025 is more than four times this figure. The higher estimate for the year 2025 implies a tenfold increase in WOCA installed nuclear

TABLE 11.1 Long-term Nuclear-Power Growth Estimates for WOCA (net GWe, by the end of the year stated)

Year	Low-High
1995	412-425
2000	504-558
2005	608-773
2010	736-1,091
2015	888-1,483
2020	1,049-1,917
2025	1,209-2,370

Source: OECD Nuclear Energy Agency and the International Atomic Energy Agency, *Uranium Resources, Production and Demand,* OECD, Paris, December 1983.

capacity over the next four decades. This is a prospect that appears remote indeed. It should also be noted that the low estimate for 1995 is considerably higher than warranted by the current state of the market. A figure of 350 GWe appears to be a more realistic estimate.

Translating the data of Table 11.1 into projections of uranium demand entails assumptions to be made regarding reactor mix and the introduction of more fuel-efficient nuclear-power reactors. A number of alternative strategies present themselves with respect to choice of the appropriate fuel cycle. Clearly, different nations may favor different fuel cycles, based not only upon the economics of power generation and security of uranium supply, but also upon the political aspects of nonproliferation, energy security, and public opinion. The OECD/IAEA has produced projections of WOCA uranium requirements based upon two reference reactor strategies and two mixed reactor strategies as regards alternative fuel cycle choices. At the upper and lower extremes of consumption requirements are 100 percent once-through LWRs and 100 percent FBRs, respectively. These are called "reference" strategies since the true mix will involve a combination of a number of alternative reactor types and not simply one extreme or the other.

Estimates of WOCA annual and cumulative natural uranium requirements based upon the projections of installed nuclear capacity given in Table 11.1 and upon these four reactor strategies are given in Table 11.2.

Reference Strategy 1 assumes that the nuclear-power market will continue to be dominated by Light Water Reactors on a once-through (i.e., "throwaway") cycle. All LWRs built in the period 1991–2000 are assumed to be of 15 percent improved type; after 2000, improved technology LWRs are installed exclusively (additions and replacements); pre-1991 LWRs are retrofitted by the year 2001. It is assumed that there will be no reprocessing to provide the plutonium used by FBRs.

Reference Strategy 2 assumes that, beginning in the year 2001, the FBR commissioning rate is determined by plutonium availability, the growth in nuclear-power demand, and an introduction rate which is such that all new reactors (additions and capacity replacements) built in 2010 are FBRs. Other than FBRs, only LWRs of the current technology type on a once-through cycle are introduced after the year 2000. Reprocessing capacity is assumed not to be a constraint on the FBR buildup.

For **Mixed Strategy 1**, an improved LWR strategy is followed in North America while the remainder of the OECD continues to

TABLE 11.2 WOCA Uranium Requirements for Illustrative Fuel Cycle Strategies (thousand tons U_3O_8)

		LWR Once-through Reference Strategy 1		Mixed Strategy 1		Mixed Strategy 2		Oxide-fueled LMFBR Reference Strategy 2	
		Low	High	Low	High	Low	High	Low	High
1985	Annual	56		56		56		56	
	Cumulative[1]	195		195		195		195	
1990	Annual	70	72	70	72	70	72	70	72
	Cumulative[1]	511	514	511	514	511	514	511	514
2000	Annual	75	79	75	79	81	83	86	91
	Cumulative[1]	875	891	875	891	886	901	898	920
2005	Annual	109	155	96	130	73	90	65	70
	Cumulative[1]	1,772	2,010	1,741	1,947	1,750	1,899	1,776	1,908
2010	Annual	131	215	111	169	66	94	59	74
	Cumulative[1]	2,374	2,933	2,258	2,695	2,098	2,358	2,085	2,270
2015	Annual	159	278	129	226	66	140	51	113
	Cumulative[1]	3,099	4,165	2,856	3,683	2,430	2,943	2,358	2,738
2020	Annual	182	345	146	278	88	1,651	59	121
	Cumulative[1]	3,951	5,721	3,543	4,944	2,817	3,708	2,631	3,323
2025	Annual	207	417	163	335	85	183	49	111
	Cumulative[1]	4,922	7,626	4,312	6,478	3,249	4,579	2,902	3,901

[1] Cumulative from 1982 (inclusive).

Source: OECD Nuclear Energy Agency and the International Atomic Energy Agency, *Uranium Resources, Production and Demand,* OECD, Paris, December 1983.

rely mainly on the LWR with some LMFBRs. In **Mixed Strategy 2**, all OECD countries except Canada follow the LMFBR strategy. In both these strategies, HWRs are used in Canada and also in some developing countries.

Supply

WOCA's uranium reserves and resources were discussed at length in Chapters 4 and 6. As exploration activity and geological knowledge expand with time, resource estimates will undoubtedly grow, although it is possible that the size of the speculative resources category may decline as well as increase. Estimating future levels of uranium exploration expenditurc, however, is an extremely hazardous task, involving not only estimates of future uranium market conditions, but also the expected state of the markets for minerals competing for the same exploration funds. Even then, there is no guarantee that future exploration expenditure will yield a success rate comparable with that which prevailed in the past.

Estimating future levels of uranium production is also fraught with difficulties. Adequate production to sustain the nuclear-power industry in the first quarter of the twenty-first century will depend on many factors, among which the most important is the availability of markets. This, in turn, will depend on the establishment of a favorable political and economic climate, both at national and international levels.

Table 11.3 presents two sets of OECD/IAEA estimates of WOCA production capability through the year 2025 that could be supported by presently known resources (RAR and EAR-I) recoverable at cost of $50/lb U_3O_8 or less. The lower set of estimates incorporates only existing and committed production centers, while the higher set includes all four types of production centers.

DISCUSSION

By comparing the data given in Tables 11.2 and 11.3, it is apparent that annual production from existing and committed production centers will not cover estimated annual requirements beyond 1990. Production would then be required from what are now only planned or prospective centers, and such centers could be called upon to

TABLE 11.3 Long-Term Uranium Production Capability Projections (thousand tons U_3O_8)

	1985		1990		1995		2000	
Country	A	B	A	B	A	B	A	B
Argentina	468	468	156	676	156	806	156	806
Australia	4,940	4,940	3,250	4,290	3,250	6,500	3,250	6,500
Belgium	52	52	52	52	52	52	—	—
Brazil	546	546	546	2,821	546	6,201	546	5,655
Canada	14,950	15,600	15,730	17,550	12,870	18,460	8,840	15,730
France	5,070	5,070	5,070	5,070	5,070	5,070	5,070	5,070
Germany (FRG)	52	52	52	52	—	—	—	—
Italy	0	0	0	0	309	309	309	309
Morocco	0	0	0	1,092	0	1,820	0	2,340
Portugal	150	150	221	663	221	663	221	663
South Africa	8,224	9,446	8,402	10,872	7,953	13,854	7,638	16,549
Spain	195	195	982	1,567	982	1,892	1,892	0
United States	13,520	13,520	15,860	17,030	18,200	24,310	14,040	23,400
Others	12,492	13,130	13,844	19,760	16,042	18,200	0	11,050
total	60,659	63,169	64,165	81,495	65,651	98,137	41,962	89,964

A = existing and committed centers; B = existing, committed, planned, and prospective centers.

meet as much as half of the annual requirements by the end of the century. Thus, the oversupply situation, which has been a feature of the uranium industry since the inception of a commercial market in the late 1960s, appears to be reaching an end, and a substantial effort will have to be put into exploration and the establishment of production facilities to ensure that the supply of uranium is adequate to meet long-term requirements.

Knowledge of WOCA's undiscovered resources (EAR-II and SR) is fragmentary. Nevertheless, it is possible that production from presently undiscovered resources could, given the appropriate incentives, cover any likely requirements for uranium during the period to 2025. The question of incentive, however, is paramount in ensuring that the exploration efforts will be sufficient to ensure the

2005		2010		2015		2020		2025	
A	B	A	B	A	B	A	B	A	B
0	780	0	910	0	1,040	0	0	0	0
3,250	6,500	3,250	6,500	3,250	6,500	3,250	6,500	3,250	6,500
—	—	—	—	—	—	—	—	—	—
546	5,655	546	5,655	546	5,655	546	5,655	546	5,655
6,500	10,530	2,990	7,020	1,560	4,160	910	2,150	0	1,560
5,070	5,070	5,070	5,070	2,600	2,600	0	0	0	0
—	—	—	—	—	—	—	—	—	—
309	309	0	0	0	0	0	0	0	0
0	2,860	0	2,860	0	2,860	0	2,860	0	2,860
221	221	0	0	0	0	0	0	0	0
7,638	16,731	6,070	15,094	5,408	12,697	3,645	11,609	2,033	7,756
0	0	0	0	0	0	0	0	0	0
12,220	26,000	9,360	17,680	5,070	11,050	3,250	6,760	3,250	4,810
0	7,800	0	7,800	0	2,600	0	1,300	0	1,300
35,754	82,456	27,286	68,589	18,434	49,162	11,601	37,934	9,079	30,441

Source: OECD Nuclear Energy Agency and the International Atomic Energy Agency, *Uranium Resources, Production and Demand,* OECD, Paris, December 1983.

discovery and development of new deposits. Not only must the financial returns on such an investment be sufficient to encourage further development of the uranium-mining industry, but also the political and social environment must be conducive to the support of any planned expansion.

NOTE

1. The methodology and results of this exercise are reported in: OECD Nuclear Energy Agency, *Nuclear Energy and Its Fuel Cycle: Prospects to 2025*, OECD, Paris, 1982.

Bibliography

Ahmed, S. Basheer. *Nuclear Fuel and Energy Policy*. Lexington Books, D. C. Heath, Lexington, Mass., 1979.

Amr, Asad T., Jack Golden, Robert P. Ouellette, and Paul N. Cheremisinoff. *Energy Systems in the United States*. Marcel Dekker, New York, 1981.

Banks, Ferdinand E. *Resources and Energy*. Lexington Books, D. C. Heath, Lexington, Mass., 1983.

Battey, C. G. "Australian Uranium Resources." *Atomic Energy in Australia* 21 (October 1978):2–9.

Beach, Charles M., and James G. Mackinnon. "A Maximum Likelihood Procedure for Regression with Autocorrelated Errors." *Econometrica* 46 (January 1978):51–58.

Bickel, Lennard. *The Deadly Element*. Macmillan, New York, 1979.

Charles River Associates. *Long-run Price Projections for Uranium: An Evaluation of Methodologies*. U.S. Department of Energy, Grand Junction, Colo., 1981.

Chow, Brian G. *The Liquid Metal Fast Breeder Reactor: An Economic Analysis*. National Energy Study, 8, American Enterprise Institute for Public Policy Research, 1975.

Chow, Brian G. "Comparative Economics of the Breeder and Light Water Reactor." *Energy Policy* 8 (December 1980):293–307.

Crowson, Philip. "Economic Factors and the Uranium Market." In *Uranium and Nuclear Energy: 1980,* The Uranium Institute, Westbury House, London, 1981.

Darmayan, Ph. "The Economics of Uranium Supply and Demand." *International Atomic Energy Agency Bulletin* 23 (June 1981):3–7.

Duderstadt, James J. *Nuclear Power*. Marcel Dekker, New York, 1979.

Eden, Richard, Michael Posner, Richard Bending, Edmund Crouch, and Joe Stanislaw. *Energy Economics: Growth, Resources and Policies*. Cambridge University Press, Cambridge, England, 1981.

Eklund, Sigvard. "Nuclear Power Development—the Challenge of the 1980s." *International Atomic Energy Agency Bulletin* 23 (September 1981):8–18.

Fareeduddin, Syed, and J. Hirling. "The Radioactive Waste Management Conference." *International Atomic Energy Agency Bulletin* 25 (December 1983):3–26.

Flowers, Sir Brian. *Royal Commission on Environmental Pollution—Sixth Report: Nuclear Power and the Environment.* HMSO, London, 1976.

Geddes, William P. "The Uranium and Nuclear Industries in China." *Resources Policy* 9 (December 1983):235–51.

Goldschmidt, Bertrand. *The Atomic Complex.* American Nuclear Society, La Grange Park, Ill., 1982.

Howe, Charles W. *Natual Resource Economics.* Wiley, New York, 1979.

International Atomic Energy Agency. *International Nuclear Fuel Cycle Evaluation, Summary Volume.* IAEA, Vienna, 1980a.

International Atomic Energy Agency. *Fuel and Heavy Water Availability.* Report of the International Nuclear Fuel Cycle Evaluation, Working Group 1, IAEA, Vienna, 1980b.

International Atomic Energy Agency. *Guidebook on the Introduction of Nuclear Power.* IAEA, Vienna, 1982a.

International Atomic Energy Agency. *Nuclear Power, The Environment and Man.* IAEA, Vienna, 1982b.

Joskow, Paul L. "Commercial Impossibility, the Uranium Market and the Westinghouse Case." *The Journal of Legal Studies.* VI (January 1977):119–76.

Koutoubi, Sani, and Ludwig W. Koch. "Uranium in Niger." In: *Uranium and Nuclear Energy,* Mining Journal Books, London, 1980.

Lecraw, Donald J. "The Uranium Cartel: An Interim Report." *The Business Quarterly* 42 (Winter 1977):76–84.

Lecraw, Donald J. "Economic Rationale for Canada's Future Uranium Policy." *Resources Policy* 5 (September 1979):208–16.

McIntyre, Hugh C. *Uranium, Nuclear Power, and Canada–U.S. Energy Relations.* C. D. Howe Research Institute (Montreal, Quebec) and National Planning Association (Washington, D.C.), 1978.

Miller, Saunders. *The Economics of Nuclear and Coal Power.* Praeger, New York, 1976.

Moore, Thomas G. *Uranium Enrichment and Public Policy.* AEI–Hoover Policy Studies, 25, American Enterprise Institute for Public Policy Research, 1978.

OECD International Energy Agency and Nuclear Energy Agency. *Nuclear Energy Prospects to 2000.* OECD, Paris, 1982.

OECD Nuclear Energy Agency. *Safety of the Nuclear Fuel Cycle.* OECD, Paris, May 1981.

OECD Nuclear Energy Agency. *Nuclear Energy and Its Fuel Cycle: Prospects to 2025.* OECD, Paris, 1982.

OECD Nuclear Energy Agency. *The Costs of Generating Electricity in Nuclear and Coal Fired Power Stations.* OECD, Paris, 1983.

OECD Nuclear Energy Agency and the International Atomic Energy Agency. *World Uranium Potential: An International Evaluation.* OECD, Paris, December 1978.

OECD Nuclear Energy Agency and the International Atomic Energy Agency. *Uranium: Resources, Production, and Demand.* OECD, Paris, December 1979.

OECD Nuclear Energy Agency and the International Atomic Energy Agency. *Uranium: Resources, Production, and Demand.* OECD, Paris, February 1982.

OECD Nuclear Energy Agency and the International Atomic Energy Agency. *Uranium: Resources, Production, and Demand.* OECD, Paris, December 1983.

OECD Nuclear Energy Agency and the International Atomic Energy Agency. *Uranium Extraction Technology.* OECD, Paris, 1983.

Nuclear Energy Policy Study Group. *Nuclear Power Issues and Choices.* Ballinger, Cambridge, Mass., 1977.

Pan Jie-Jian. "Distribution of Non-circular Serial Correlation Coefficients." *Selected Translations in Mathematical Statistics and Probability* 7 (1968):281–91.

Quirk, James P. *Intermediate Microeconomics.* Science Research Associates, Chicago, 1976.

Radetzki, Marian. *Uranium: A Strategic Source of Energy*. Croom Helm, London, 1981.

Ranger Environmental Enquiry. Australian Government Publishing Service, Canberra. *First Report,* 1976. *Second Report,* 1977.

Rybalchenko, Igor L., and J. P. Colton. "Spent Fuel Management." *International Atomic Energy Agency Bulletin* 23 (June 1981):36–40.

Taylor, June H., and Michael D. Yokell. *Yellowcake: The International Uranium Cartel*. Pergamon Policy Studies, Pergamon Press, New York, 1979.

United States Department of Commerce. *The Commodity Shortages of 1973–74*. Washington, D.C., 1976.

United States Department of Energy. *Survey of United States Uranium Marketing Activity*. Energy Information Administration, Washington, D.C., annual publication.

United States Department of Energy. *Statistical Data of the Uranium Industry*. Grand Junction, Colo., annual publication.

United States Department of Energy. *Projected Costs of Electricity from Nuclear and Coal-fired Power Plants*. Energy Information Administration, Washington, D.C., Vol. 1, 1982.

Uranium Institute. *The Balance of Supply and Demand, 1978–1990*. Mining Journal Books, London, 1979.

Uranium Institute. *The Uranium Equation, The Balance of Supply and Demand. 1980–1995,* Mining Journal Books, London, 1981.

Warnecke, Steven J. *Uranium, Nonproliferation and Energy Security*. The Atlantic Institute for International Affairs, Paris, November 1979.

Zimmerman, Charles F. *Uranium Resources on Federal Lands*. Lexington Books, D.C. Heath, Lexington, Mass., 1979.

Zorn, Steven A. "Recent Coal and Uranium Investment Agreements in the Third World." *Natural Resources Forum* 6 (October 1982):345–57.

Index

About the Author

ANTHONY DAVID OWEN is a senior lecturer in econometrics at the University of New South Wales. Until 1974 he was lecturer in economics at the University of Bradford Management Centre. During the academic year 1984–85 he was senior research fellow in the School of Economic Studies at the University of Leeds.

Dr. Owen has published widely in the areas of econometric theory and natural resource economics.

Dr. Owen holds a B.A. from the University of Leicester and an M.A. and Ph.D. from the University of Kent at Canterbury.